Enzymes
for Autism and other
Neurological Conditions

The Practical Guide for Digestive Enzymes
and Better Behavior

———

Enzymes
for Digestive Health
and Nutritional Wealth

The Practical Guide for Digestive Enzymes

———

A Story, An Adventure,
Practical Information and Science

———

Karen DeFelice

Published by ThunderSnow Interactive
Fourth Printing with revisions and corrections 2005
Third Printing (Second Edition) with revisions and corrections 2003
Second Printing with revisions and corrections 2003
First Printing, Technical Review England and USA 2002
Printed and Bound Minnesota USA

Enzymes for Autism and other Neurological Conditions
The Practical Guide for Digestive Health and Better Behavior
ISBN 0-9725918-7-7

Enzymes for Digestive Health and Nutritional Wealth
The Practical Guide for Digestive Enzymes
ISBN 0-9725918-6-9

Cover photo by Michael S. DeFelice
Cover designs by Karen DeFelice
Illustrations by Karen and Michael DeFelice

Contents

Acknowledgements

I would like to thank the following individuals who have assisted me in many different ways in this work. Of prime importance is Cindy Kelley, co-moderator of the Enzymes and Autism group. Cindy's dedication helped so many people and ensured the forum runs so smoothly, truly a support and backbone of the group. Since April 2001, Cindy has contributed greatly helping evaluate situations and determine a practical and appropriate course of action. Besides being a devoted mother of two boys, one with an autism spectrum condition, she has a Master's degree and practical experience in special education. My deep appreciation to her husband Mitch as well, because of all his support keeping the effort going and encouraging us all.

A tremendous thanks to all the reviewers, especially the individuals doing the technical and scientific reviews, and those that contributed ideas, information, and direction ensuring the information was accurate to the best of our ability, and improving the quality of this work.

Thanks especially to all the parents and members of the Enzymes and Autism group who contribute their time, effort, insight, and experiences evaluating the use of enzymes as part of their child's or their own program. The drive of each and every person fighting the battle now bursts forth as the phoenix rising showering a cascade of hope and health. The devotion to helping our children and ourselves with the blessings of our Lord brings the results for not only our families, but others still on this journey, and just starting out. This book is dedicated to all the parents, families, caregivers, and others caring for people they love with neurological conditions. And to those many blessed children and adults who deal with and live with these conditions every day. This is a story about hope.

Thanks to all who made this venture a reality. The information, ideas, and science are due to many, often volunteers. To the Feingold

Association, who reviewed the chapter on 'Sulfur, Epsom Salts, and Phenols' ensuring the text was accurate and filling the gaps appropriately. And for all their extensive research into the matter of food additives, biological, and behavioral relationships. To those individuals who provided great insight on many other matters.

My deep appreciation to Pastor John Kline, who simply does his job very well. I can only express it as his support, his words, his actions were important, and it made a difference. Thanks to our school nurse, who has worked with our family with a cooperative spirit and openness to make my sons' days at school far easier enabling them to get the most out of their education.

Extraordinary thanks to Dr Devin Houston, who had the vision for and took the tremendous risk in formulating specific enzymes for the autism spectrum conditions and now others, and who diligently supported the parents and other adults who chose to try enzyme therapy in helping their children and themselves. Many thanks for Dr Houston's patience and assistance in teaching about enzymes, and providing a practical alternative. History is filled with those incredible moments when a unique individual does an amazing thing that changes lives for the better. We are all deeply indebted.

To my dear husband Dr Michael DeFelice and my two wonderful sons Matthew and Jordan for just being the profound blessings of my life. Especially for Mike's patience, support, guidance, and love all these years.

And thanks to our Great Lord God
through Whom all things are made possible.

Introduction

'Many neurological conditions, including autism, may be improved with digestive enzymes. Is this a new special breakthrough?'

Not really. The excitement is not due to a brand new scientific discovery. Enzymes are not new, and have existed since before the dinosaurs. The big dinosaurs had to have enzymes to digest the little dinosaurs, and the little dinosaurs needed them to digest the plants they ate. So, enzymes themselves are not new. Enzymes are a natural and essential part of all living things. They are catalysts, which means they speed up biochemical reactions without being changed in the process.

'What about enzyme therapy? That's got to be new. I've never heard of it before.'

Enzyme therapy is not new either, and has been in use for ages. In the 1700s, Jean Senebier identified that gastric juices could be applied to wounds to speed healing and reduce infections. The enzymes in these juices would degrade the dead tissue and keep the wounds clean. Plant extracts were used before that for the same reasons. Enzyme therapy has been around for some time, but many times, unfortunately, the association with other elements and philosophies diminished the fundamental biochemistry involved. This has led many to see enzyme therapy as less than reputable. Currently, enzyme therapy enjoys common usage with sports injuries, pancreatitis, cancer, fibromyalgia, and viruses, among other things.

'Okay, so what's the story? Don't tell me, a brilliant scientist (who should be dashing and full of charm in this story, by the way)

concocted something remarkable in his lab involving enzymes to help with the problematic symptoms of autism. He gave his formulation freely to all the needy and there was much rejoicing.'

Not exactly. Well, this story does involve a smart, thoughtful, and devoted professor-researcher type person as part of the tale, and he does come up with a 'breakthrough' enzyme formulation, but that is only part of it. Then the real movers and shakers came on the scene.

'Okay, I got it…there was a beautiful young maiden with long golden hair, of course, who was desperately ill and the scientist's formulation saved her. They married, lived happily ever after, and then they gave these enzymes freely to all the needy and there was much rejoicing.'

That's a good ending, but not the one here. In our story, a group of parents and other concerned adults working tirelessly to help themselves and their children from an assortment of disabling problems latched onto the scientist's enzymes. The field of enzyme therapy was 'rediscovered,' only this time applied to autism spectrum and other neurological conditions. The children were improving immensely in a matter of weeks; in some cases, even in a few days. Success stories with enzymes were popping up like dandelions in the spring.

'So what is so new about all this? Seems like it has all been done before…perhaps a new twist on an old story?'

Enzymes are not a new silver bullet miracle for all people. But certain new enzyme formulations seem to be providing a major short-cut through the current chaotic maze of health alternatives. What's new is that recent science is revealing a wealth of medical aspects to autism, attention deficit, sensory integration, and related neurological conditions. Now that these are identified, it is only logical to look at what measures are effective in improving these conditions. Supplying digestive enzymes orally is speeding up the process of improvements, and making other therapies much more effective. Enzyme therapy appears to be the next step and ties many of the other pieces together. Enzymes remain catalysts after all!

It turns out that a great many individuals with behavioral and neurological problems have some fundamental disorders arising from their digestive systems. And enzymes are in an excellent position to be on the front lines in digestion. This book covers why enzymes may help these children and adults. The text is not based on theories that stretch the imagination or require a great leap of faith, rather it shows and relies on very fundamental, known biochemical principles supported by a wealth of practical research and experience.

The discussion here shows the types of improvements currently available with digestive enzymes and how they are already used with real people in everyday life. The results and guidelines of this experience are presented here. This is for information only and not to be taken as medical advice for individual situations.

Although enzyme therapy has been around a long, long time, not too many books are out on it, and the ones available tend to fall into a few limited areas. Some are strictly biochemical ones, like those used for physiology and chemistry. Others follow much older theories with some parts since proven true while other parts...well, need updating. Another feature of most books is that they do not address the practical everyday use of digestive enzymes.

People have different opinions on what constitutes 'recovery.' Some believe if you no longer qualify for the diagnostic criteria, you no longer have AD(H)D, autism, Asperger's, or whatever. This makes sense. However, because other people feel that being on the autism spectrum is a matter of being neurologically wired a certain way, you never really recover. They feel that some individuals just reasonably learn how to control their behavior so it is socially acceptable. These people say the individual is not really changed in neurological wiring but has just adapted well, and 'hides' the unwelcome manifestations of his condition. Therefore, the person is not 'recovered' but rather has learned coping strategies.

This book does not go into that area or argument. Nor does it consider the many behavioral therapies available. Behavioral therapies and biological ones work together for a total program. This book concerns some of the biological conditions that often accompany a diagnosis. The references of 'recovery' made here are in relation to a person being improved in physical health and reducing some of the suffering brought on by the imbalances in biological function. Many individuals may need additional therapy and assistance besides the biological concerns. It is important to understand that no one therapy will help everyone. Or no two things, or five things.

Book Organization - Read what you like

This book is grouped into descriptions of individual experiences, practical information, technical explanations, and reference sections.

If you are interested in the personal stories, look at Chapters 1, 7, 12, 19, 20, and 21. Also, the very first section and last few paragraphs of most chapters contain a personal experience.

If you are interested in just reading the facts, science, and technical aspects of enzymes, see Chapters 2, 3, 4, 5, 6, 8, 9, 10, 13, 14, 15 and 19. Also, look for sections in most chapters marked 'The Science Behind It,' usually the second section in each chapter.

There is an extensive reference section to support all the ideas and information contained herein. The Reference section includes the references indicated in the text. Further References lists helpful books. The Even Further References includes other supporting information and related references not specifically noted in the text.

If you are interested in the practical tips and suggestions, these are scattered throughout the book. However, Chapters 6, 9, 11, 16, 17, 20, and 21 as well as Appendix A have more emphasis on this.

Preface to the Second Edition

Enzymes and digestion problems are similar in many people no matter which behavioral, physical, or medical diagnosis you have. This edition incorporates and reflects some of the more recent research and information since previous printings. The Contents has been expanded. A new section, Chapter 9 – Why You Should Eat Like a Pig, was included to provide insight into how enzymes have been used extensively with animals and veterinary science for many decades with wonderful results. Chapters 17 and 18 focus more on food chemicals both naturally occurring and added, and overall detoxification. Please see the Disclaimer at the end of this book.

A note for readers

The intention of this book is to assist anyone even remotely interested in enzymes or neurological conditions. Because I am a parent, and a great number of the people I associate with are parents, and the majority of people who participated in this adventure with me are parents dealing with helping their children, I tend to write 'parents.' However, this is not meant to exclude non-parents. Definitely, all non-parents are included. I struggled with how to word things appropriately without having to resort to a multitude of clunky sentence structures or distant terms such as 'caretakers of people in healing.' I decided on 'parents and other adults' and this is meant to mean anyone giving or taking enzymes for any reason. Even if it only reads 'parents,' everyone else is included. An 'other adult' includes aunts, friends of the parents, grandparents, adults with autism conditions, adults with digestive issues, adults with autoimmune conditions, adults caring for other adults, doctors, therapists, child development professionals, researchers, school nurses, teachers, specialists in the field, students, teenagers, really bored people who just happened to pick this book up by accident, and anyone else who tends to breathe air.

I appeal to your graciousness and ask you to please overlook the fact I was not more creative in coming up with a better way to refer to you if you are not a parent. You are very important, and I sincerely hope that you are able to find something useful here from my journey through the maze. Think on what you can use and enjoy the rest. Hopefully, this will encourage and help others in some way.

P.S. If you have any questions or concerns for me, here is my contact information. I will try to be helpful.

kjorn@thundersnow.com
5720 Wentworth
Johnston, IA 50131 USA

CHAPTER 1

The Two Little Princes

Once upon a time, there was a very typical not-so-average family. This is the story of my family. Although it really starts before, this particular journey starts with me. The best way to put it is just to say I had a very hard time growing up for a number of reasons. The adventure became even more colorful during my teenage years when I developed an eating disorder, depression, and non-stop migraines. Or maybe the non-stop head pain led to the depression and chewing problem. Any way you look at it, it wasn't pleasant. I have had constant migraines and severe sensory issues all my life, but no one talked about 'sensory issues' back then. I saw many doctors and tried lots of therapies. I spent lots of money, but didn't get anywhere. The specialists told me I was depressed and stressed out, try to relax, and basically just 'learn to live with it.' I really worked at it too, because I had the ultimate goal of feeling better and moving on, which never happened.

I eventually met a tall, dark handsome professor and we married. His name is Mike. He is a very wonderful person.

The much more interesting part starts when our first son was born. He was remarkable. Blond hair, blue eyes, and desperately cute. The 'Little Prince' of our lives. However, it became very apparent that our Little Prince was also very...*unique*. Mike described him as 'high maintenance.' The best adjective for his overall disposition was 'intense.' Little Prince Matthew was intense about everything. He nursed intensely, he ate intensely, he played intensely, he even crawled intensely. His crawl style attracted a lot of attention. He crawled like a charging bull, with his knees off the ground, bottom in the air. He would look at something, put his head down, and charge at it. More than once we were sure he was going to smash right into a wall or table, when at the last second, only inches from a head-on collision with a firmly rooted object, he would abruptly switch direction and continue crawling furiously on.

And that wasn't the only thing that would switch abruptly. We could set his moods by the clock. He would be happily playing or riding around doing errands for two hours, and then break down in fits of screaming no matter where we were or what we were doing. Or he could be wailing away then suddenly get interested in something and dart off to inspect it. This intensity continued as a toddler. Matthew rarely watched television. When he did, it was more like he was vigorously memorizing or analyzing it rather than enjoying it. He was keenly interested in gadgets, gears, and mechanical things all the way down to knowing which stores had matching doorstops and all the different types of hinges used. He did point at objects he was interested in. However, one of the only things he was interested in was identifying burned-out light bulbs. He always let us know, loudly, whenever he spotted a burned-out light that needed replacing.

Matthew is extremely bright, and a wonderful artist – in that almost disturbingly gifted sort of way. He could draw with uncanny depth, perception, and detail, but he just could not cope or function with day-to-day events, especially outside our house.

Matthew seemed to have a radar that detected anything within eight inches around any part of his head. He would jerk away if anything came closer. It became quite a challenge to brush his teeth, wash or cut his hair, wash his face, or pull any clothing over his head.

Transitioning was bad, too. Did I say bad? I meant horrendous. If we planned everything out quite well and no unexpected changes popped up whatsoever, we *might* get through the day with only a couple of meltdowns. This included going to sleep. I could put the happy little tyke in his crib with a kiss good night. Within a minute, he would be screaming himself to sleep. You would think someone had snuck in and lopped off a limb. Five minutes later (I would time it on the kitchen clock), he would be fast asleep. The longest five minutes of the day.

He frequently screamed inconsolably. Only Mike could calm him by rocking with him in a chair chanting 'Matthew and Daddy rocking in the chair. Rock, rock, rock. Rock around the clock' over and over again. And rock Matthew did. The rhythmic, unrelenting round the clock rocking, bouncing, and head-banging was his trademark.

He had the trademark behaviors of constant rocking, round the clock head-banging, non-responsiveness to people or animals, wouldn't be held, the awkward motor control, obsessing with gadgets, extreme sensitivity to his environment, and he cried and screamed a lot. These characteristics and more were all clearly part of him before he started the series of successive antibiotics for ear infections between the ages of one and two. I kept asking what was wrong and why Matthew would get another ear infection as soon as the previous one had cleared up! I used to joke that we should just buy a five-gallon jug of amoxicillin antibiotic in bulk from a discount warehouse at the rate we were using it.

It wasn't like I didn't ask around or badger the doctors about what was wrong or what I could do. But it was always the same; nobody seemed to think much about it.

'You are just a nervous first time mother. It just seems unusual to you.'

'You are overreacting.'

'Toddlers often act like that.'

'He would have different symptoms if it was allergies. Not that.'

'Most kids get lots of ear infections. Don't be alarmed.'

'Head-banging is very common for about 15 percent of boys. He will grow out of it.'

But he did not grow out of it. Or out of his need for a rigid schedule. Or his aversion to affection and touch. In fact, he only got worse.

Jordan was born when Matthew was 21 months old. Our second Little Prince was also very different. Very quiet, affectionate, and easy-going. We really enjoyed this change of pace. He didn't have the chronic ear infections either. Or at least he didn't show it. As a matter of fact, he didn't seem to show much sensitivity to any pain at all. Our pet name for him was 'Jumpin' Jordan,' due to his tendency to stand so quietly for so long and then just take a flying leap off the furniture…or into the furniture. This was an alarming skill for his parents to watch, but since it didn't seem to bother him, even when he smacked hard into something, we chalked it up as his own unique learning style. Now that I know more, I realize this is a different form of sensory integration problems. He was too *in*sensitive to typical sensations and physical sensory inputs.

Jordan never had the physical screaming tantrums his brother showed. He didn't have the inability to transition. And, although the rocking was there, it was not nearly as intense or compulsive. I wouldn't have thought there was any problem at all with Jordan except for the mounting gastrointestinal problems he developed, which his brother did not. He had escalating chronic constipation alternating with diarrhea, picky eating, chronic fatigue, gagging, lethargy, bacteria infection, and constant headaches. Jordan was like me in that he couldn't stand sunlight, bright lights, or even moderately loud noises. Jordan even looked a lot like me. Brown hair, facial features…all the way down to the constant dark circles under his eyes!

Matthew continued to be so completely miserable; it was obvious something was wrong. At three years old, I took him for an official evaluation through a local children's habilitation center. I asked about possible allergies. The specialists dismissed allergy tests because 'he didn't have the right symptoms.' I didn't know about the difference in being intolerant of a food or chemical instead of being truly allergic to it. We periodically took one thing out of his diet at a time to see if he was sensitive to it, and never saw any change. They did behavioral observations, hearing tests, eye tests, and some other things. Everything seemed okay, but yet it wasn't.

What about autism? I brought up the possibility a few times. What about the incessant rocking and gadgets and other behaviors? The lack of affection and eye contact and social skills? What about that streak of giftedness? What about his rigid structure and the strict logical thinking? I wasn't concerned about what we labeled the problem; I only wanted to get on with the business of healing it, if possible.

Since Matthew could speak quite well, passed the hearing tests, and was very creative, they felt classic autism wasn't *it*. The evaluation team said it was something very much like autism but not quite; or he was 'gifted' in creativity and had severe social problems; or maybe extreme ADD with severe depression and let's just throw in some chronic anxiety attacks for good measure (okay, she didn't really say that last part that way – I was getting frustrated). A woman doing one of the evaluations said, 'He is a little too odd to be autistic.' Now, is that better or worse? I was a little relieved it wasn't *that*, but disappointed that I did not have any better clue as to what to *do* about the situation. So, I needed to keep looking for a solution. I didn't know that there was a 'spectrum' with all these personal characteristics and possibilities. Now I understand it would be something like Pervasive Developmental Disorder (PDD) and severe Sensory Integration Dysfunction (SID).

We resigned ourselves to the fact that there was a neurological condition and we would do the best we could to help our boys. I suppose we very appropriately had the unofficial diagnosis of:

PDD-NOS = Pediatrician Didn't Decide – Neurologically Odd Son

Since several people on both sides of our family have trouble with depression and anxiety, and I had constant migraines and nerve sensitivity problems, it seemed like a logical explanation. None of the doctors had any other recommendations. Matthew grew older. We worked and played and tried to teach him coping skills. And we really enjoyed his uniqueness. Wow, was he *unique*.

Matthew only socialized with his brother and parents. In preschool, he hardly spoke a word, or even moved much for that matter, but he ranted in terrible fits at home. He was irritable, whiny, and wailing in a very distraught way – like a wounded wild animal. He hated to change schedule. He could not get himself dressed in the morning without me to help prod him along. He would cry when he had to focus, read, or do much mental work on command (he read far better at home than at school). We lived through rivers of tears. He would cry frantically for Yellow Blankie, his blanket, instead of Mom or Dad. He was hypersensitive to the air and everything around him. He quickly became overstimulated whenever we went out of the house and he needed to rest very frequently. He gnawed on everything (our furniture, his clothes, straws, whatever). He had awkward physical movements – okay, downright uncoordinated. He was very afraid to do physical things such as riding a trike, going up a slide, scared of the plastic ball pits at the fast-food restaurants. His sensory problems were seriously debilitating.

I want to take a break here from all the depressing parts to say that there were many good parts as well. We had a lot of fun, too. Mike and I loved the boys totally and admired all their unique abilities. Matthew would act out complete replicas of television shows, including the commercials, copying all the right voices and terms. Deep plodding voice: 'Our documentary on turtle habitats in the wild will continue after these messages.' Switch to bouncing animated voice: 'Our peanut butter is the peanut-iest. Buy it today for your family. Your kids will love it.'

Matthew and Jordan were best friends, simply inseparable. I very much believe that having each other helped them both from possibly withdrawing ever more. A bridge keeping them connected to other things. Having two of them was both fun and helpful. They were at the same interest level. Matthew was the type of kid I could hand a straw to, and he would keep both himself and Jordan entertained with jokes and puppet shows and silliness for a good 20 minutes. When the straw lost its luster, I could hand him a paper cup next, which would be good for another 20 minutes of fun. Incredible. His imagination was both so very deep and wide. However, it only extended to Jordan, and then to his parents.

Jordan had an easier time of it. He was snuggly, cuddly, and cute. Besides Matthew, his dearest friend was Little Bear. I had brought this floppy stuffed bear home for Jordan's first Christmas when he was six weeks old. Little Bear stayed connected to Jordan for around the next five years. Simply glued to Jordan. Jordan would hold Little Bear up to his face in one hand and suck his thumb on the other hand. Little Bear wore a long green ribbon around his neck, and Jordan would twirl the ribbon round his fingers. And twirl, and twirl, and twirl. At least it was a very quiet repetitive behavior.

Basically, we worked with the boys a great deal at home and maintained a fairly low-key but active structure. I became the primary 'therapist' and taught them at home. We did a lot of one-on-one interaction, and what I know now as floortime play. We did a great deal of story-telling, books, reading, and computer games. We acted out dramas, having stuffed animals reenact shows we saw on TV. There was water play, mixing kitchen ingredients for 'experiments,' crafts, visits to the local science center, and the sorts of things moms like to do with their kids. Letting them explore the world at their own pace (in a structured way) was a big part of my style. Mike would spend hours upon hours with them building very complex structures (for that age) out of Lego blocks and other toys.

By the time Matthew was five years old, I got so exasperated with the difficulties and knew Matthew was not 'growing out of it' that I insisted he at least try the antidepressant Zoloft for any depression and anxiety. This helped. He was better on Zoloft but far from good, and was starting to have serious problems as he entered public school. His first grade teacher was concerned he would never be able to write on a straight line. His second grade teacher at first quarter conference 'failed' him in everything, even though he was already in remedial groups. She said he could barely recognize words although she thought he was bright enough. I knew he could read at home, but he fell apart just dealing with the school day. He was constantly lying down on the floor whenever they had group time (headaches, too much stimulus). And he had another characteristic that was particularly unhelpful in a classroom.

Matthew thinks in a very literal way. At times this is wonderful. At times it is a hurdle. His kindergarten teacher and first grade teacher really worked with him over and above the call of duty. I was impressed with how much effort they gave helping my son. But sometimes it was me who finally figured out what he needed. Probably because, most of my life, many caring and thoughtful people, and some rather unthoughtful ones, have pointed out that I also tend to see things very literally. I never thought of this characteristic as being either good or bad. It just was a particular characteristic I had. It wasn't until I had a son of my own who thought the same way that I realized how downright irritating it can be at times.

In first grade, Matthew needed to work on handwriting his letters appropriately. He would not write on the lines as his teacher showed him. So, after talking with his teacher, I tried to figure out the problem.

'Matthew, do you know your teacher wants you to write your words on the lines?'

Matthew answered, 'Yes.'

'So, why don't you write on the lines?'

'She didn't say to.'

'But you know she wants you to.'

'But she didn't say to do it.' Matthew was very sincere about this.

Think…what is going on…?

'Matthew, write on the lines whether she tells you to or not. When someone says they *want* you to do something, that means to *do* it.'

The next day, he was writing on the lines.
The day after that, he had not written on the lines.

'Matthew, why didn't you write on the lines?'

'I did. Yesterday.'

Think…what is the connection he needs…?

'Matthew, write on the lines every day. When your teacher teaches you something, she means for you to continue doing it all the time afterward as well, whether she tells you to or not.'

The next day, Matthew's paper had lines and writing all over it, going up the margins, looping around the top, everywhere.

'Matthew, what is this? Why aren't you writing on the lines?'

'I did. I just decided to draw my own lines and write on those.'

Arrrggggh!

> 'Matthew, write your words on the lines on the paper and only
> on the lines that are on the paper until you graduate from any
> school you ever attend, whether anyone ever tells you to or not.'

We haven't had any problem since…with lines. However, I admire
his creativity.

This is just one example of many where I needed to intercept
and figure out how a literal mind is hearing and translating 'typical'
expectations and directions. Being Matthew's mom has helped me to
understand how frustrated people would get with me at times. They
must have felt I was being difficult on purpose, thinking, 'how could
someone not 'get' something so obvious,' when I sincerely didn't
understand their meaning. I am not sure why Matthew's abilities to
be so vastly creative and yet, at times, so brutally literal reside so
nicely together. It appears inconsistent, yet underneath, somewhere,
perhaps there is an explanation we just don't yet see.

It seems to me that everyone has a 'blind spot' in learning and
understanding things. Many people don't *get* algebra or chemistry.
How many people just laugh off the fact they can't program their
VCR? These are deficiencies you can usually work your life around
or completely avoid. It is much harder when your blind spot happens
to be reading social cues…something you can't get around with other
people. It is right there in your face, every day. However, at one
time or another everyone has this problem with fundamental
communication and making themselves understood.

During this time, we also worked with a child psychiatrist and
psychologist at a children's hospital. We made picture charts for
communication and organization, we did special clothes, we had
reward and reinforcer systems, positive feedback, negative feedback,
zero feedback. Nothing worked. For two years, I tried persuading
the specialists to try other medicines, therapies, anything else. Finally,
they were going to develop an Individual Education Plan (IEP) for
Matthew, and we would officially be on the road to specialized education.

Then, something really remarkable happened.

Beige: A Splash of Color
– Living with constant migraine pain

When Matthew was seven, I went to a new neurologist for my own constant head pain. I have lived with constant migraines...and I mean constant – 24 hours a day, seven days a week – all day and all night for as long as I can remember. A relentless pain that never let up. Sometimes I could not function for days; constant upset stomach and chronic fatigue. Long ago, starting when I was a teenager, I was told I was depressed and stressed. The advice was to just get to the root of my 'suppressed emotions,' then I would be fine, and all my neurological issues would go just away. I have had almost every tooth in my head replaced because of the constant gnawing on stuff to deaden the pain. You might call this my repetitive behavior. Twenty years of therapy later and not much better, I got to the point of having tried everything else. I was not imagining this. It was real pain. As in Owwww! My head hurt, plain and simple.

My best attempt at describing this pain: it is as if my head was full of broken glass. Thousands of tiny shards stuck in my brain. Sitting completely still hurt, but moving hurt too because it caused the shards to shift. Another description: it is like my brain is soaking in acid. A hot, stingy feeling. Like when you scrape your leg and need to put some antiseptic on it. It has the initial stinging feeling. A burning sensation. I was usually very nauseous. Lights and glare and loud noises and bright colors were all irritating. I was so hypersensitive to everything. A sensory integration disaster. I could only wear certain clothing textures for a certain amount of time; needed sunglasses on any non-cloudy day; and I keenly noticed any two-degree change in temperature. Unless you are in the arctic about to freeze to death and need to know the exact temperature, this is not a useful skill.

I was very limited in the type and amount of light, sound, temperature, colors, food, and people I could tolerate. I still cannot figure out what the color orange is for! That has got to be the most irritating thing ever invented. Now beige is great color. I find it to be a wonderful accent to white, light grays, or taupe. I would work intensely about five hours a day, then collapse. I ended up as a computer programmer, technical writer, and developing educational print materials and software from home. This allowed me to at least be in a controlled environment where I could work at my own capacity. This may explain why many people with neurological difficulties need rigid structure. It was easy for me to provide the structure Matthew and Jordan thrived best in because I desperately needed it myself.

In my late 30s, I eventually found a neurologist who evaluated my situation. She listened as I told the story once more of the constant pain I can never get away from, even during sleep. The kind of pain in your head that is so ingrained in your life and has been there so long you can't tell where the pain stops and you begin. She nodded, and almost whispering repeated, 'Yes, I see. Yes.' Later I found out that she herself suffered terrible migraines, although not constantly. She considered my condition to be real pain and a real neurological issue, saying they would classify it as chronic daily migraine. This appears to be one of those catchall categories when there is no other place to put you.

She said I had an inherent neurological condition. She described it as saying my nervous system was ratcheted up about 260 percent higher than most people. I just took in way more sensory information than most. Also, the nerves in the brain become inflamed and pressure develops for biological reasons. She said the exact mechanisms are not precisely understood and it would take a variety of measures to manage it. This included medication, diet, lifestyle changes, and some alternative therapies. She explained about reducing the total load on the nervous system, and that included looking at diet, sleep, chemicals, and anything that 'stressed' the body. Some of the pain and difficulties could be relieved but the basic neurology would still be there. She very concisely summed up everything I have come across since then about neurological conditions, including the autism spectrum.

Mike sighed, 'Great! I married a woman who really DOES have a headache every night!'

The neurologist prescribed a medication called amitryptiline. It helped immensely. Wow, did I feel a lot better! The pain receded; I could sleep soundly through the night. My visual perception changed. What I mean is that it took about a month for me to adjust to seeing life differently, literally. Especially in depth perception. Like when you get new glasses. I kept commenting to my husband, 'Everything looks so *flat* now.' The best way to describe it was that things used to look more like a hologram with the glowy, 3D effect. I know, it is a little tough to explain. Now, it is like just pleasantly being in life instead of having to fight through a 12-inch thick wall of pain to see and deal with everything.

I immediately insisted that we try Matthew with this medicine. Our neurologist agreed to see him and started him on it. Matthew immediately improved in three to four days. When he started amitryptiline, he *really* turned around. He was in second grade and his reading score had been 30 words per minute in the second quarter. This was about as close to the bottom of his grade level as you could get. By the third quarter, just four months later, it was 130 words per minute. WOW! His temper calmed, he socialized, he became more attentive in class. This was the type of improvement you see happening to other people on television living in far-away places. This was eye-popping, jaw-dropping, can-you-believe-THAT-type improvement.

Matthew slept all night, he dressed himself, he stopped gnawing his cloths to shreds, and academically he soared. But, at home, he still crashed. He still cried, screamed, argued, and could not transition well. Medication helped a lot with the sensory issues. Before this, I had this nightmare of the teenage years looming ever closer as we were stuck in the mud. Time was marching on with us in tow and the situation never improving. Now, there was hope that a brighter future lay ahead.

During this time, younger brother Jordan had many of the same multiple sensory sensitivities, although not as severely and not as debilitating. He had a lot of physical stomach and bowel problems instead of the tantruming and screaming. He would just linger around the house, flopping over furniture, saying he was 'so tired' and had 'a bad headache.' He always wanted someone to hold him while he sucked his thumb, holding Little Bear and twirling the green ribbon. Jordan did not eat well. He was getting very thin for his age and not gaining much weight at all. At six, he was still having excessive bowel problems. We started him on the medication too, and he improved somewhat all around.

My neurologist suggested taking magnesium as well for calming and pain, and to look for any 'triggers.' Triggers can be anything that provokes the headaches: food, chemicals, loud noises, crowds, even changes in the weather.

'The weather!' I wailed. 'There is no way to keep the weather constant!'

'That's what makes this so challenging,' she replied.

She very strongly advocated a regular sleep cycle. Go to sleep at the same time and get up at the same time. The 'same time' being within 15 minutes. Apparently, your nervous system can get really out of whack if you do not maintain a regular sleep/wake cycle. People with sensitive nervous systems are very susceptible to this. She was right. By staying up an extra hour or more, I definitely could feel it the next day. I was also much better on cloudy days than sunny ones; much better on cool days than hot ones.

I tried a few other medications and therapies attempting to improve more. One was a prescription medicine called baclofen, which is a muscle relaxer often given for spasms and neurological dysfunctions. The idea was to relax the muscles that were pressing on the nerves, thus adding to the constant pain. The baclofen really helped in the beginning and I took it regularly for about a year. However, eventually I found it was not really improving the situation any more, or maybe I had learned to relax the muscles myself. I was tired of being slightly drowsy all the time, too. I still use it occasionally when muscle spasms kept me awake at night.

Overall, I would say we improved about…50 percent, just about the amount my neurologist had told me in the beginning. We were rather stuck at this level now. Everything was much better, but there were still enough bad days. Life was now just 'incredibly difficult' instead of 'extremely disastrous.' And Jordan started getting worse. Another year went by.

Some of the known triggers for migraine are wheat, chocolate, cheese, bananas, chemical additives, and other common foods. Food elimination may not fix the total problem, but any relief at this point was welcome. So, we decided to try removing dairy first. While talking with my sister-in-law about living without milk, she casually mentioned that she and her sister are both lactose intolerant. So is their mom. What?! After ten years, I find this out! Most of my husband's immediate family are lactose intolerant!

> 'Oh, yes, we can't have any milk at all or it brings on terrible headaches and digestive problems. Maybe Jordan is lactose intolerant and that is causing his headaches and gut problems!'

Dearly as I did not want to leave dairy behind, I did consider that there were special types of milk in the store just for those who were lactose intolerant. This wasn't going to be so hard after all. So, we took out milk and dairy products for a week. No change. So, I put them back in for a week. Nothing. Took them out for another week. Nope. Put them back in. No difference. I tried eliminating other foods as well but nothing improved. Drat.

While continuing to look into possible triggers and therapy, I found some information on sensory integration dysfunction at the bookstore...by accident. I was walking by the shelf and there was a book facing outward called *The Out-of-Sync Child*.

> 'Yeah, boy,' I snickered to myself as I sauntered over to look at it, 'I got a couple of those!'

I flipped through the book. Imagine my surprise to see it describe us perfectly! I was mumbling to myself as I scanned the pages faster and faster, ever more intensely:

> 'Yep, Matthew does this.'
> 'Wow, I do that!'
> 'Ooooo, this describes Jordan perfectly!'

However, as fascinating a find as all this was, I was far more interested in what to *do* to bring this nightmare to an end. And this led to another very interesting turn of events.

Sensory integration and migraines – The science behind it

Sensory integration is the process of the body compiling all of the external stimuli and signals from the environment, analyzing it, organizing it, and integrating it. With sensory integration dysfunction, one's nervous system is not able to properly sort out and make sense of all the sensory input that the senses are taking in. Sometimes, the central nervous system cannot process the information in a typical or appropriate manner. Something may be hampering the central nervous system in some way, or the nervous system is just unable to process all the information received. Many children with pervasive developmental disorders also have a wealth of sensory processing difficulties. It is not clear where one issue ends and another one starts.

All that confusion and stimuli coming into the brain may trigger migraine pain. Having too much total 'load' for the body to handle can also make you very susceptible to migraines. Most people recognize that migraines exist in adults, but may not realize migraines also occur in children. Ample research exists showing the prevalence of childhood migraine and its related effects (Egger 1989; Wendorff 1999). Migraine is often associated with many other issues such as food intolerances, seizures, epilepsy, hyperactivity, environmental sensitivities, and other symptoms (Carter 1995). As a small footnote, I was evaluated for possibly having epilepsy about nine years earlier because of pseudo-seizure muscle spasms.

Once you identify that migraine or other pain exists in children, it is easy to see why there may be such a wealth of sensory integration problems. Living with migraines is known to make one hypersensitive to sound, lights, color, temperature, and other things. Even motion can make you feel ill. Motion such as climbing up and down steps. Or having to go shopping when you are trying to rest. Motion requires moving your aching head around and usually means changes in light and temperature. If a child is sensitive to this type of motion, and therefore resists readily moving from place to place, an adult may describe them as 'having difficulties with transitioning.' Transitioning problems might be due to pain or confusion, and not just the person simply being difficult. We found out later that Jordan gets motion sickness quite easily, even from riding the school bus or in a car. This was one source of his headaches and wanting to be carried a lot.

Sometimes the person needs less sensory input and sometimes they need more. Not getting this input leaves a sensation best described as floating, detached, anxious, or confused. As if the sensory system is looking for something that it needs before it can go on; like it needs an anchor. The necessary input might be in the form of deep pressure, which helps many children. The nervous system is seeking more pressure in order to organize itself and calm down. I need to wear textured nubby socks made from cotton or wool. Slick, dress socks drive me batty. The texture of the thick socks provides more sensory input on my feet. I don't know why this is necessary, but I get to the point I can't function unless I have the right type of socks on. I also cannot sleep without several blankets on my feet. It was not until I learned about sensory integration that I realized this is because I needed the additional pressure on that spot. Without it, I toss and turn all night and wake up with a headache. I bet many people have these little 'quirks' they rarely confess.

That pressure helps people to calm down is not so unusual. This may be why hugging someone feels so good. Sometimes you can be so neurologically hypersensitive that you need something different than just a typical hug. Sensory issues are more a matter of degree. What is a 'loud' noise to me, is only moderately loud to someone else. I think a wool sweater on my skin feels good and cozy whereas someone else may think it is too scratchy to wear. The total load of sensations for one person may feel like they are dealing with an attacking tiger, whereas someone else senses a little fluffy kitten.

Another way to help focus the nervous system might be through abrupt sensations to help achieve the feeling of being anchored. Some kids throw themselves into furniture. Others may scratch themselves.

Being both oversensitive and undersensitive to a typical environment can cause many problems in daily functioning. A lot of the quirky behaviors, preferences, rituals, and rigidness shown by people with certain conditions may just be a logical way of trying to cope with a particular environment. Many of the repetitive or self-stimulatory (stimming) behaviors seen may be a method of trying to analyze, organize, and integrate sensory information. Repetitive rocking or focusing on something can be ways of calming the nervous system while you, or it, tries to make sense of things. As the person is doing the repetitive behavior, they may be screening out excess sensory information, taking in other information, and trying to make sense of it all. Many adults typically have things they like to do to 'unwind' after a hard day. These rituals help us to regroup within ourselves.

As I understand it, amitryptiline helps to induce the deep sleep that some people cannot achieve on their own. This deep-sleep cycle plays a part in the regulation of pain sensitivity. So, taking this medication reduced some of the hypersensitivity we were experiencing. Amitryptiline has other pain-control properties as well.

The adjustments in sensations and pain interpretation went a long way in diminishing all the sensory integration problems we were having. It was wonderful! Now, a light breeze is barely noticeable instead of feeling like bugs crawling all over my skin. Everyday sounds stay in the background instead of sounding like a series of explosions. I still question the creation of the color orange, though, but at least it doesn't make me nauseous anymore. Just getting a good night's sleep is incredibly rewarding to the body. Imagine how it feels to constantly have to function on too little sleep night after night after night.

When Matthew started amitryptiline, the invisible barrier concerning activities around his head disappeared – haircuts, brushing teeth, brushing hair, washing hair, and wearing hats. All of that is okay now. He still was not affectionate in the least, and days were difficult. However, he had traded in being non-functional, and advanced to barely-functioning-badly. Both boys said the medication greatly reduced their headaches.

The *Out-of-Sync Child* is an excellent book on sensory integration. It does a wonderful job of describing what is going on inside the person with sensory dysfunction compared to the behavior you see on the outside; and compares how a person with sensory integration difficulties sees the same situation as a person considered neuro-typical (NT). This book gives many exercises for helping strengthen sensory integration abilities as well as ideas on how to better manage the individual's environment. Specialists have specific training in sensory integration therapy, and this usually is in the realm of an occupational therapist (OT). Some sources of information on sensory integration therapy are given at the end of this book.

I found I had already been doing much of the appropriate sensory therapy anyway by default. Most of the exercises are not difficult, not expensive and you can do them at home yourself. Knowing what is going on with sensory processes and what to do about it can avoid many unpleasant situations, and gain a happier family. The nervous system can recover and rebuild to a certain extent.

Sensory therapy - Re-training the nerves

Nerves communicate with one another through electrical and chemical signals. Neurotransmitters are the messengers. Nerves have an inherent certain plasticity or flexibility. Nerves respond to stimulation. They can grow and 'learn' by repetitive stimulation, and build new response systems and patterns. Targeted repeated stimulation is the basis of many sensory integration exercises. Nerves can learn to respond to certain types of stimulation, and new neural pathways can be built.

However, repetitive messages of pain from a given location may establish a trained response we may not want in the same way. Usually, pain is a very normal and beneficial mechanism, often warning us when something is wrong or our body is being harmed. At times though, the pain response is not appropriate to the actual situation.

There are different types of pain. You may experience different symptoms based on which type is involved, or how the pain reaction is 'remembered.' Inflammation, trauma, infection, temperature, or a multitude of other things can activate pain. If the pain triggering event is sustained or repeated, the body may develop a 'pain memory.'

The pain reaction may be such that the body responds with a magnification of the event. This means that a very small sensation is registered as a huge sensory input. The nervous system 'over-reacts.' An example would be if the first event or two was eating a lot of crunchy food resulting in cramping of the stomach and gagging, the nerves may unfortunately learn this response. Afterwards, a very small amount of that crunchy food, or any type of crunchy food, may cause the same huge cramping and gagging reaction. In some people, the system might under-react (hypo-sensitive).

Another possibility is that the body remembers the pain even though the initial event causing the pain is no longer there. The pain response is 'stuck.' Even with the actual pain source gone, the nerves still react like it is there (example: the phantom pain experienced by some after having traumatic surgery).

When the nerves in the gut experience a distorted sensory response, it can lead to digestive and eating problems. Often kids who are picky eaters may turn out to have strong sensory reactions related to their digestive tract. They may strongly prefer crunchy foods, or crave creamy foods, or refuse to eat cold foods, or gag up every time they try to eat soft foods, or become irritable if brightly-colored foods are on the plate. The reaction may not be related to the makeup of the food, but the texture or smell may be triggering physical reactions in the gut. Some fragrances may provoke a very real headache, or nausea. Similar reactions may happen with sounds.

The pain memory is part of irritable bowel syndrome for many people. There may have been a difficult bout with constipation, or bacteria infection, or inflammation in the bowel. The original problem may no longer be there, but the bowel remains extra sensitive. Similar to injuring a joint playing sports, then the joint heals from all outward appearances and test results, but from then on the joint remains extra sensitive, and 'acts up' from time to time.

Sometimes there may simply be a genetic component where the nerves are inherently just more sensitive and no particular traumatic event ever happened.

So, does the inappropriate response stay stuck forever? Hopefully not. The system can un-learn the pain response which is no longer needed, or even appropriate, on its own with time. Sensory integration therapy may facilate this process. Some medications are beneficial as well. There may be other reasons sensory integration works depending on the therapy used, but realizing this aspect of pain, or nerve unresponsiveness, and re-training the nervous system can go a long way in improving the situation.

Martial arts

One of our strategies for helping with sensory issues was to enroll the boys in some type of martial arts program. There is a good fit between the characteristics of this sport and the needs of those with sensory issues. Martial arts help with orienting your body in space, muscle strength, coordination, balance, discipline, and focus, among other things.

You do this sport as an individual, but also participate in a group. You can do it all year round in a controlled environment without worrying about the weather. Solid structure and repetition help you slowly build skill upon skill as you train your body. Building self-esteem, self-control, and confidence are cornerstone benefits. It is also an appropriate place to yell, kick, jump, and punch if a person needs or wants to do that. Many parents attest to how martial arts really brought out improvements in their child with sensory issues.

Each school and branch of martial arts has its own personality and focus, so do visit several to find one appropriate for you or your child. We chose one particular Taekwondo program because there was only light contact at selected times, it emphasized defense, had a strong program for growing children, and the program focused on the character-building benefits rather than just attaining belt levels.

Mike pointed out that kids who are shyer and not as socially savvy are often targets for teasing and bullying. We hoped Taekwondo would provide some guidance and coping skills for our sons as well as build their physical skills. A feature that made this program work for us was they had a well-structured, high-quality program, yet a very flexible schedule. If needed, we could attend class later that week, if the boys (or the mom) were having a particularly over-whelming day.

Matthew started at age six and it was a struggle, but worth it. Orange belt was the third level where Matthew missed two promotions and was working towards the third attempt. One of the members joked that he would outgrow that belt before he promoted out of it. Although the instructors were deeply committed to all the students, deep down, I was afraid there was a bit of truth in that. Matthew started the amitryptiline shortly after that. His classmates and instructors all noticed his sudden change in attitude and effort for the better . . . and the mom was very pleased to move on to a different color belt. It was still a struggle, but he was definitely keeping up now. Although we hit our fair share of dead ends in helping our boys, Taekwondo ended up being one of our best and wisest decisions.

Being aware of the sensory impact of activities and environments can significantly improve results with highly sensitive individuals.

CHAPTER 2

Why Enzymes?
– The Gut-Brain Connection

In learning about sensory integration, migraine, and how foods influenced behavior, I quickly discovered that a lot of what is known about alternative therapies for childhood neurology is found among parents dealing with conditions of the autism spectrum and attention deficit. Maybe it is because these 'spectrum' conditions are like a catchall for the many people who do not fit neatly into some other category. Neurological diagnoses are often overlapping, complex, vague, and subject to the personal background of the person doing the evaluation. A neurologist may identify certain aspects and give you one diagnosis. A psychologist may give you another from a behavioral point of view. A family practitioner may give you still a different one from a medical perspective. These may all be parts of the same whole, all interconnecting. I was back looking at autism once again.

The available relevant information wanders through how the gut is intricately connected with the immune system, the nervous system, behavior, and learning. The symptoms and behaviors produced are all interwoven. Recent research has uncovered many newer physiological connections. One book, titled *The Second Brain* by Dr Michael Gershon, outlines how the majority of the serotonin neurotransmitters were found to be located in the gut (as opposed to only in the brain). Other neurotransmitters lie there as well. In fact, every known neurotransmitter present in the brain is also present in the gut. Thus, there is a real physical basis to having a 'gut reaction.'

In addition, there are many gastrointestinal issues that commonly plague those with neurological conditions. This is not terribly surprising once you think about it. However, it is very nice to have scientific physiological reasons at hand, and not be left with just 'poor parenting,' 'you're over-reacting,' or 'it's your imagination.'

As interesting as all this was, my main concern was what to *do* about it. I wanted Jordan's gut to be functioning well, and Matthew to be just plain functional in life. My personal goal for myself was to be able to wake up in the morning without feeling as if I was just run over by a truck with parts left rattling in my head. You know, to be able to just get up, and get dressed, and do really average things in a really average day, and go to sleep without a ripping headache. I wanted to know what practical steps I needed to take in order to improve whatever was possible, and how to manage the rest of it.

Elusive diagnoses, variable treatments

Most neurological conditions are not one specific thing but a diagnosis that encompasses a range of possible conditions, symptoms, and causes. The current criteria for diagnosing a person with something such as autism or related condition are based solely on behavior and noticeable outward symptoms, not on biological or medical test results. This means that if an individual exhibits a certain number of predefined behaviors then he qualifies for a diagnosis of autism, Aspergers, attention deficit, or something else. It is called a spectrum because of the myriad of possible symptoms found in individuals and the range in how extensively these symptoms may be expressed.

Two different people can receive the same diagnosis yet show completely different behaviors, tendencies, and reactions. They may also have very different underlying biological issues. Because of this, specialists use a wide variety of therapies. One combination of therapy may work wonders for one person while a very different set of methods is essential to bring relief to someone else.

No one has identified a single known cause or even two or three known causes of these conditions. Current thinking believes it is some combination of genetics and environment, which can occur in many different combinations. Neurological conditions have a strong tendency to run in families, although they can also surface randomly. Equally at play are the environmental factors. Some parents have related their child's difficulties or specific symptoms to particular stressors, certain foods or chemicals, physical illness, or some event. In other cases, such as ours, there is no one particular moment when symptoms started. While a certain food, stressor, or chemical may not directly be the cause of a condition, it may greatly aggravate symptoms. Research is on-going in many areas.

The main characteristics of a person dealing with these spectrum conditions follow. Not every one exhibits all of these.

- visual disturbances, avoiding eye contact
- sensory integration issues: light, sound, taste, touch, textures
- lack of affection
- lack of social skills or interest in surroundings
- withdrawn, anxious, afraid, or hyperactive for no apparent reason
- eating, digestion, and bowel difficulties
- general learning disabilities, inattention, distractibility
- lack of appropriate verbal communication
- fixation on certain rituals or objects, or repetitive behavior

The thing to remember is that any of these behaviors could be caused by a number of problems. And little children may not be able to express what they are feeling as an adult would. A toddler or preschooler is very unlikely to say, "Mommy, I am experiencing a migraine with acute throbbing over my right temporal lobe." He is much more likely to tantrum, not eat, and bang his head on the floor. Even adults have problems indentifying what is actually ailing them.

Why enzymes for autism?

Since autism was first formally identified as a special condition all its own in the 1930s, people traditionally considered it an area of mental health. Later, it passed into the realm of neurology. In the past few decades, increasing evidence has surfaced showing that many individuals diagnosed as being on the autism spectrum suffer from a variety of physiological issues that need attention on the biological level. These concerns can involve the nervous system, gastrointestinal system, and immune system, and result in many of the behaviors and tendencies that fit into the official diagnostic criteria. The amounts of information showing a physiological connection are increasing in several areas. It could also be that some professionals are not looking immediately at possible medical problems and giving appropriate treatment. They may be looking at behavior and diagnose on that basis alone.

Physiological aspects affecting neurology and behavior

Symptoms can arise for a number of reasons. The digestive system is critical in this relationship, probably more so than many people realize. The digestive tract is a gateway of consumed substances coming into the body. What are the possible biological problems currently identified? The primary issues include:

- leaky gut syndrome (intestinal hyper-permeability)
- food allergies, sensitivities, and intolerances
- chemical sensitivities and toxic overload
- overworked immune system
- sulfation dysfunction
- various nutrient deficiencies
- yeast, bacteria, parasite infections or overgrowth
- viral infection
- metabolic dysfunctions

Several of these conditions are well researched in themselves. (And exploring into possible treatments, research also finds digestive enzymes may be effective regarding each area). You may recognize many of these individual issues as more and more prevalent in the general population. Someone may not have a particular diagnosis, but you may know people who:

- 'are allergic to everything'

- 'are always tired, moody, and irritable'
- 'are sooooo forgetful and inattentive'
- 'show general disruptive behavior'
- 'are sooooo sensitive'
- 'are always sick with something'
- 'have undefined fatigue, aches, and other "problems"'

This further explains why each person needs to be treated as the individual they are, and why so many different programs might help at least some. The underlying biology varies.

Think of it like going to a pizza shop. Everyone there is having pizza. Everyone's pizza has some similarities, like a crust, some sauce, and toppings. Everyone recognizes a certain pizza aroma and atmosphere, too. However, each person may have a very different type of pizza. One person may have half of his or her pizza of symptoms attributed to yeast overgrowth. This leads to a fourth of the pizza devoted to the resulting leaky gut and the last quarter going to effects from food intolerances. Yeast treatments and a yeast diet may go a long way to restoring this person's health. In addition, he or she may take certain supplements for supporting digestive health – enzymes, probiotics, and zinc. This combination may be all this person needs for outstanding relief from adverse symptoms.

Another child may have a pizza which has five out of eight slices topped with symptoms resulting from accumulated toxins. We then see two slices attributed to digestive dysfunction, and one slice due to bacterial overgrowth. This person may not see the success the first person saw with an antifungal to control yeast. This child may see the most success with digestive support and a detoxification program.

A third individual may have a more assorted pizza. This biological pizza sports one slice for constipation, which led to another slice of bacteria overgrowth. There is also one slice of inflamed intestines, creating one slice of malabsorption, leading to two slices of specific nutrient deficiencies. Because of this, he or she also ends up with a slice of overburdened liver, and one slice of chemical sensitivities in the body. This one is tougher. Neither antifungals nor removing toxins may tackle a major portion of the problem here. This person may need a much more comprehensive approach to restoring good health.

In addition, many people take the option of going to the salad bar of behavioral therapies or medications as part of the overall plan. Each family assessing its own situation and going with what works best for its members is critically important. You will find very sound reasons why a therapy that 'works for everyone' may not give positive results for you.

For my family, Matthew had many of the behavior issues, whereas Jordan and I had obvious digestive problems. All three of us were racked with migraines. We had an assortment of symptoms even within our own family, even with the same environment, same menu, and same lifestyle. It is interesting to note that even though my older son did not show any typical signs of digestive problems he responded quite well to certain enzymes. We seem to be very representative of the array of situations possible with neurological conditions, some apparent and some not-so-obvious.

It is fairly easy to see why some of the primary underlying biological problems are related to one another, yet others don't look like they refer to the same issue at all. Are all of these things connected? Or do we just pick one and see if it applies to our situation? This is what interested me when I started reviewing all the possible connections. I was interested in finding the most efficient pathway through this chaotic maze. As I told my husband in the beginning, it appears to be a giant free-for-all with many theories all going in different directions. Parents were running here and there trying this supplement or that therapy, with the general direction of 'Just try it and see what happens. Good luck!' Upon closer inspection, maybe this is not such a big rambling hodge-podge after all.

Just think of all the diets that come and go for losing weight. We have the high-protein diets where you focus on proteins, balance your vitamins and as a token line, remember to drink eight to ten glasses of pure water each day. Or the carbohydrate diets where you focus on complex carbohydrates, eat smaller and more frequent meals, and drink a healthy amount of pure water. Or consume 80 percent raw foods, cook sparingly with no additives, and supplement with water to move the fiber along. Or the liquid diet, which allows you all the liquids you want until you float away. Or [insert favorite diet here]. If you step back and look at all these diets, you start to notice that one of the things they all have in common are the eight to ten glasses of pure water. So, maybe it is the *water* that is really helping you to achieve an appropriate weight. Maybe the *water* is a major factor in weight control instead of having a minor role.

If we look at all the possibilities of the conditions noted above, are there any common players? In reading through tons of material, in filtering through most of the alternatives, there is a member of the team usually in the background having a big impact, but not prominently noted as a major star – the fundamental digestive enzymes. It is very evident from much of the research that many people with neurologically related conditions may have a multitude of gastrointestinal problems. And because digestive enzymes are key players in the gastrointestinal tract, they are on the front lines of what happens there. Specialists may not have unveiled the connection or come around to considering the possibility of these enzymes… until now!

Digestive enzymes are notoriously effective in facilitating and supporting intestinal function, immune system health, detoxification, and many other things. Enzymes may improve the digestion, absorption, and utilization of whatever food you eat, or supplements and medicines you take. The person taking enzymes may just feel so much better that he or she is in a better frame of mind to learn, think, work, and interact. In this way, enzymes may contribute to behavioral therapies being far more effective. In essence, digestive enzymes may be a fundamental short-cut to better health and learning. They remain the great catalysts after all!

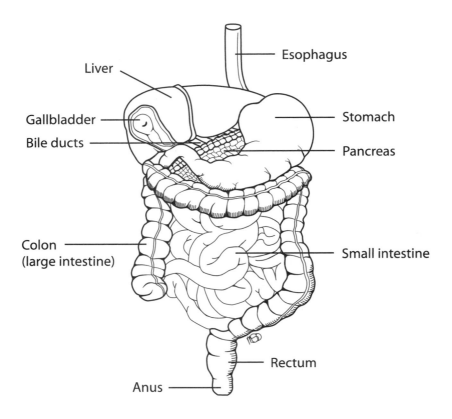

Liver

Gallbladder

Bile ducts

Colon
(large intestine)

Anus

Esophagus

Stomach

Pancreas

Small intestine

Rectum

Figure 3.1

Major Parts of the
Digestive System

CHAPTER 3

Digestive System Basics
– Getting to the Bottom of It

It may not be that 'you are what you eat' or even 'you are what you digest' but really 'you are what you absorb' – or even what you do not absorb. Many with gastrointestinal problems may not only have trouble digesting food and supplements but also might not be able to absorb anything they do digest. Taking supplements by the handfuls may not be doing much good if the body is not absorbing, transporting, and using those nutrients. All those supplements might be passing right on through doing nothing at all. Food and supplements must be digested and absorbed to be of use to the body.

First, let us define what we are talking about as far as digestion is concerned (refer to Figure 3.1). Usually, talking about the 'gut' refers to any and everything from the first passage of food or drink past your lips all the way through and out the other end of your body at the anus. At the beginning, you have your mouth, followed by the esophagus, which is basically your throat. The esophagus is very important but not quite as exciting as the other parts of the digestive system. The esophagus connects to the stomach.

The stomach is located under the rib cage and heart on the left side of your body, and is connected to the small intestine. Sometimes people refer to 'intestines' as encompassing all your guts, or other times as everything below the stomach. After the small intestine comes the large intestine, also called the colon. At the end of the colon is a part called the rectum. Then the final point of departure is the anus. The gastrointestinal system or digestive tract refers to this entire system, usually excluding the mouth and esophagus. This will be the meaning for these terms used in this book unless otherwise stated.

Now did we leave anything out? Oh, yes, where do the liver, pancreas, and gall bladder go? Since the stomach is on the left side of the body above the intestines, we will just locate the liver on the right side, and add the other organs tucked in around them for now. All of this is held together with some very strong muscles and skin. That's basically how it looks.

Now, how does it work? The digestive tract is quite a very dynamic environment and is in constant motion physically and biochemically. What happens there directly impacts every other process in your body.

The brain – Food for thought

Digestion actually starts before any food passes your lips. Just the thought of eating, the sight of food, or the smell of something cooking can trigger the brain to start the process of welcoming food into the body. The brain starts in motion the process of getting enzymes, digestive juices, hormones, and saliva ready for action. Your metabolism may increase in anticipation, revving up for action.

A very complex biochemical process is taking place. The entire process follows a well-ordered sequence of events. You have direct control over two parts of the process. You decide what goes in the mouth including how long to chew it, and usually when to allow things out the other end. Everything else happening in between functions automatically.

Digestion is under control of the brain and nervous system. However, you can assist this process by a variety of things. Digestion goes much smoother if you eat slowly and in a relaxed manner, have a well-balanced diet, chew your food thoroughly, and eat something about every four hours. Drinking plenty of pure water throughout the day may help. Digestion impacts everything else that happens in your body, as we shall see.

The mouth – Mechanical muncher

Starting at the top, you chew your food and it mixes with the saliva in the mouth. Saliva contains an enzyme called amylase, which starts to break down carbohydrates. By chewing your food longer and more thoroughly, the food is ground up into much smaller pieces before it hits the stomach. This increases the surface area of the food. Greater surface area provides more sites of contact for enzymes, which makes them more efficient in breaking down the food. Also, chewing longer mixes the amylase enzymes in your saliva more thoroughly with the food before it leaves the mouth. Eating slower allows you to enjoy tasting and chewing your food leading to greater satisfaction. Eating slower also helps prevent overeating, and that helps keep your body weight in the right range.

Once you finish chewing and swallow, the food moves down through the esophagus to the stomach. Well-chewed food moves through the esophagus quickly and easily. At the bottom of the esophagus is a little door called the esophageal sphincter. It lets the chewed food through but then closes so any stomach contents cannot easily get back up. Sometimes this door doesn't function well as in cases of heartburn or gastric reflux.

The stomach – Biological blender

In the stomach, any additional enzymes already in the food kick into gear if the food is raw. Cooked food will not contain active enzymes. The amylase mixed in with the food from the saliva is also still at work here. When food enters the stomach, the brain receives a signal that food is present and it is time for digestion to begin. The brain sends a signal to the pancreas that there is food in the stomach and digestive enzymes are needed. The brain also signals the liver to produce bile.

So now we have the digestive function in motion, with food sitting in the stomach, and the pancreas and liver preparing their digestive solutions. If digestive enzymes are taken orally and derived from plant or microbe sources these enzymes will start breaking down the food in the stomach, along with any enzymes present from raw foods and any amylase added from the mouth. Enzymes derived from animal sources usually are not active until later when they reach the small intestine. Plant- or microbe-derived enzymes can be breaking down food for up to 90 minutes before animal enzymes even get started.

Think of the stomach as divided into two parts, the upper stomach and the lower stomach, based on the function in each area. The stomach is similar to a blender. The upper part of the stomach acts most like a holding container, like the upper part of a blender. Food remains in the upper part of the stomach approximately 45 to 60 minutes (perhaps longer at times) digesting as much as possible. Then it moves to the lower stomach where muscles churn and grind and pump the food together, like the blades whirling in the bottom part of a blender.

When you put a lot of solid food in a blender, then turn it on, the blades whip up the food in the bottom part first turning it to mush. But the food sitting in the top of the blender may remain only slightly affected. So you open the lid, stir the food up a bit, then turn the blender on again. Whatever is in the bottom is mushed the most. Since we can't reach down our throat and stir our food in the gut, we need a way to drain off the processed mush in the bottom part so the food in the upper part can move on down and be processed. Fortunately, we have such a trap door. At the base of the lower stomach, a muscle regulates the passage of food out of the stomach. This muscle expands and contracts periodically releasing small amounts of the processed stomach contents into the small intestine.

When food enters the stomach, it triggers the release of stomach acid and the digestive enzyme pepsin into the stomach. Pepsin is an enzyme that breaks down proteins but is inactive until it comes into contact with the acid. The stomach acid activates the pepsin in addition to breaking down food itself. The powerful stomach acids, also called gastric juices, help dissolve the food along with the mechanical breakdown in the lower region of the stomach. A healthy person produces two to three quarts of gastric acid each day.

One of the acids produced in the stomach is hydrochloric acid. This acid kills most bacteria and other pathogens that enter the body along with any food or drink. The world is just full of bacteria, parasites, molds, and viruses wanting to reach the warm and nutrient-rich environment of the intestines. The stomach acid essentially sterilizes everything that comes down the esophagus. Low stomach acid might allow an opening for pathogens to settled in.

Well, if this stomach acid can dissolve meat and live micro-organisms, why doesn't it dissolve the stomach itself and the rest of our guts as well? There is a wonderful reason we all just don't digest ourselves away. Several layers make up the gut wall. The main one of concern here is something called the mucosa. The mucosa is a single layer of cells that moderates nutrient and liquid absorption, and also produces mucus, which covers the stomach lining. It also controls which substances will not be allowed to pass into the bloodstream. Although thin, the mucosa is very tough. In the stomach, it must withstand the stomach acid produced in addition to protein degrading enzymes. In the small intestine, the mucosa also absorbs the nutrients and water, and blocks potentially harmful elements out.

At times, this mucosal lining may become damaged. If this happens in the stomach, peptic ulcers can develop. If the mucosa in the small intestine or stomach becomes injured, the wrong type of particles may pass across and into the bloodstream causing all kinds of havoc. In addition, beneficial nutrients may not be absorbed properly. The intestinal lining replaces itself every five days or so. This is good for people with injured guts because gut healing starts immediately.

The food lasts in the lower stomach for one to two hours being churned and ground and pumped along with the enzymes and acids. Everything is just cooking away. This stomach acid also helps convert certain minerals into forms that the body can absorb. Even though the stomach is processing these minerals, it is not absorbing them. Most of the nutrients are absorbed in the small intestine. Some water, alcohol, and certain salts are absorbed right in the stomach. This is good when we need to rehydrate ourselves, but also explains why the effects of alcohol can be felt so quickly. It is soaked right up. High-fat meals stay in the stomach longer than low-fat meals. Meals containing some fat help us to feel full longer. A very low-fat meal or diet will leave you feeling hungry faster, and thus provoke your body to want to eat more again sooner.

The intestines – An action-packed drama in three parts

When the stomach has sufficiently finished its job, it turns the food into a substance called chyme, which should be the consistency of pea soup. The muscle at the base of the lower stomach releases small amounts of processed food into the intestines bit by bit. If food is not adequately broken down, larger particles will enter into the small intestine. This puts a greater burden on the rest of the system to finish the process of digestion.

The semi-fluid stomach contents start their journey through the three parts of the small intestine. The first part is called the duodenum. The duodenum is the shortest part but is the site of a lot of important activity. Remember earlier how the brain sent a signal to notify the pancreas and liver to get busy? By now, the pancreas has produced pancreatic fluid and the liver is responsible for making bile. When the stomach acid enters the small intestine and hits the gut wall of the duodenum, this triggers the layer of cells lining the duodenum (called the epithelium) to produce several hormones. One of these is called secretin. The secretin then signals the pancreas to release the pancreatic fluid into the the duodenum. This fluid contains sodium bicarbonate and digestive enzymes.

The bicarbonate raises the pH of the stomach contents because otherwise the acid coming from the stomach may injure the sensitive intestines. A less acidic environment also allows the body's digestive enzymes to work better. So, the acid created in the stomach did its job and is now neutralized in the small intestine. The pancreatic fluid also contains a number of substances important to digestion including proteases, amylases, and lipases – enzymes produced by the pancreas. These enzymes enter the intestines to help digest the food coming in from the stomach. Proteases break down proteins, amylases break down carbohydrates, and lipases break down fats.

In addition to producing enzymes and sodium bicarbonate, another important job of the pancreas is to regulate blood sugar and metabolism. When the pancreas detects that blood sugar levels are too high, it produces insulin. Insulin starts the process that moves glucose out of the blood stream and into the cells for use and storage.

This is necessary to make room for the new nutrients coming in from the digested food in the small intestine. If this mechanism is not working well, you can develop hypo- or hyper-glycemia and even diabetes.

The liver has also been busy. The liver is a virtual chemical factory where hundreds upon hundreds of reactions take place. It processes nutrients absorbed through the small intestine as well as the other drugs, chemicals, hormones and pretty much any other substance introduced into your body. As substances pass through, the liver decides what to do with them. The liver may pass a substance through as it goes on to its destination, incorporate it with other elements first into other compounds, doom it to the waste pile, or simply hold it in storage until later. The liver also cleans the blood by removing wastes and toxins, and manages over 2000 enzyme systems. If the workload becomes too much for the liver, it may not be able to process the vitamins and nutrients we take as supplements and they become wasted. It also may not be able to adequately process out all the waste products and toxins our bodies accumulate, leading to serious consequences as toxins build up in the rest of the body. Keeping the liver detoxified and clean is essential so it can function efficiently maintaining our health.

The liver also produces bile. The primary job of bile is to assist with the digestion of fats and eliminate waste products. Bile also contains bicarbonates, which help neutralize the stomach acid entering the intestines. Wonder what the gall bladder is for? Ever wonder what it actually does? The gall bladder is essentially a holding bin for bile. The liver connects to the gall bladder, which then connects to the first part of the small intestine (the duodenum). The bile produced by the liver is stored and regulated in the gall bladder until needed in the duodenum. If the bile becomes too concentrated in this holding area, a variety of elements may precipitate out. As the precipitates grow, they develop into gallstones.

Back to digestion. So now we have the pancreas and the liver producing their respective compounds and fluids. These digestive juices converge through special ducts and flow into the duodenum as the stomach contents enter the small intestine. At the same time, the small intestine is also producing and contributing some digestive fluids of its own. All of these digestive juices, bile, enzymes, bicarbonates, and processed food mix together continuing the work of food breakdown while travelling on through the small intestine. Digestion is in full swing. The food glides from the duodenum through the second and longest section of the small intestine, the jejunum.

As the stomach contents pass through the long and winding small intestine, nutrients and liquid are absorbed from the churned and digested semi-fluid mixture and into the bloodstream. The innermost layer of cells in the intestines is the epithelium cells. In the intestines, this layer of epithelium cells folds up and down forming long finger-like projections, called villi. This folding of cells increases. the surface area so more nutrients, salts, and liquid can be absorbed.

To increase the surface area further, each villus has even more tiny little projections called microvilli (sometimes referred to as the brush border). A protective mucus covers all the villi and microvilli. Think of a luxurious cotton bath towel with those little absorbent loops sticking out. Now picture each loop covered with a multitude of little hairs. Roll this up into a long yet narrow tube, and you have a model of the small intestine. (refer to Figure 3.2)

The villi lining produces some important enzymes: lactase, sucrase, maltase, and dextrinase. These are disaccharidases – enzymes which break down certain types of starches and sugars. These enzymes break down the disaccharides as these complex sugars come into contact with the microvilli brush border. The final products are water soluble monosaccharides, or simple sugars, which are absorbed immediately.

Similarly, some enzymes known as peptidases are located in the mucosa as well. These enzymes break down protein fragments, called peptides, from larger fragments into much smaller two or three part fragments. The enzymes may even break the peptide down to its component amino acids. The very small peptides and amino acids are then absorbed easily.

Another enzyme orginating from the small intestine mucosa are small amounts of intestinal lipase for splitting certain fats and fatty acids. These mucosa lining enzymes are important to note because if the mucosal gut lining becomes injured in any way then the production of these digestive enzymes is impaired.

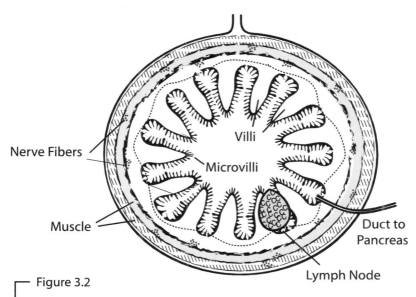

Figure 3.2

Cross Section of Small Intestine -
Nerves and Immune System (Lymphs) are Intricately
Connected throughout Digestive System

As the contents from the stomach pass through this tube, nutrients and other substances are selectively absorbed into the bloodstream as they pass the villi projections. The villi and microvilli are the ultimate door into the bloodstream. Through absorption, they selectively admit the beneficial nutrients and liquids as well as block the passage of substances determined to be potentially harmful. The intruders may be insufficiently broken down food particles, microorganism by-products, chemicals, or any other matter the villi consider unacceptable.

When this intestinal wall loses its integrity or otherwise becomes damaged, it can also lose its ability to regulate what does or does not enter the bloodstream. When a series of holes occurs in the gut, this condition is called leaky gut syndrome (because the gut contents passing through the intestines are 'leaking' past the mucosa and into the bloodstream when they should not be). Leaky gut can lead to a multitude of problems, such as migraines, malabsorption, skin problems, food intolerances and allergies, bowel problems, and a host of autoimmune conditions. (see Chapter 5 Intestinal Permeability)

This part of the intestines is where most of the vitamins, minerals and nutrients are absorbed through the intestinal wall into the blood stream. Once in the bloodstream, metabolic enzymes transport these nutrients and use them in all parts of the body, in every living cell, and in every function that makes the body work.

The ileum is the last section of the small intestine. All nutrients are absorbed by the time the contents reach the end of the ileum except for vitamin B12, which is absorbed here. The bile salts and some of the digestive enzymes are reabsorbed back into the body at this point and reused. The ileum ends where the large intestine, or colon, begins. The entire trip through the approximately 20 feet of small intestine usually takes between three and five hours. In comparison, the colon is only about four feet in length, but much larger in diameter. Contents may spend between four hours and three days in the colon before passing out of the body.

The colon – Teaming with life

The colon looks like an upside-down U. A protective mucus covers the inside of the colon, but the colon has a much smoother surface (no villi or microvilli here). What enters the colon is mostly water, elements called electrolytes, and waste products. The waste products include plant fiber, bacteria, and dead cells picked up along the digestive tract. The semi-liquid contents enter the colon and travel up the right side of your body, across the abdomen, and then down the left side, ending at the rectum. Trillions of beneficial bacteria reside in the first part of the colon ascending up your body. This is where the digestive process is completed. The helpful bacteria kill off disease-causing microbes, produce lactase, and other beneficial by-products. They also produce beneficial short-chain fatty acids and the vitamins A, B, and K.

Most of the remaining liquids are absorbed in the colon, leaving semi-solid waste that accumulates in the rectum. Stools start to form in the second part of the colon going across the abdomen. The longer the waste remains in the colon before being excreted, the more moisture is absorbed leaving a more compact and hardened mass. If the contents pass too quickly and the liquids are not sufficiently absorbed, you see diarrhea. If the waste remains too long, you have constipation. Many illnesses result when there is a disruption in the timely flow of waste through the colon, or you do not have the proper bacterial balance, or the colon is not otherwise functioning properly. These include colon cancer, ulcerative colitis, active colitis, inflammatory bowel disease, irritable bowel syndrome, Crohn's disease, parasites, and hemorrhoids. Ouch, what a list!

Transit time – Are we there yet?

To get a rough estimate of how long it takes your body to process food from one end to another, you can test for transit time. A general description of how to figure transit time follows. Take about one gram of charcoal tablets, which you can buy from a pharmacy or health food store, and record this as your start time. The charcoal will turn the stool black. You can also eat three or four whole beets, which will turn stool a deep red. When you see either the black or the red stool emerge, this is your ending time.

Calculate how many hours have passed between the starting and ending times. If it is less than 12 hours, the food is moving rather quickly through your system. You are probably not absorbing all the nutrition from your food and have malabsorption problems. From 12 to 24 hours is considered a good transit time. More than 24 hours is a bit long. With prolonged transit time, toxins can build up and start to be absorbed into the body.

The ENS – Who is minding the store?

The enteric nervous system, or ENS, regulates the passage of food and all the intricacies of the gastrointestinal system. The ENS is comparable to the central nervous system (CNS). The ENS, also called the gut brain, involves a long list of hormones and neurotransmitters that have complex effects on gut movement and fluid transport through the mucosa. It also regulates immune functions in the gut. The ENS network communicates with the brain and the central nervous system through several nerve pathways, although it operates somewhat separately from these other two systems. The nerves run all along the intestines right under the layer of villi. Whatever happens to the gut, impacts the nerves in the mucosal lining. Part of this is due to physical location, part due to biochemical reactions in the gut environment. And whenever something trips the nerves, the nerves send signals to the brain, either appropriately or inappropriately. This is how the gut, the nerves, and the brain are intricately connected.

The digestive tract needs to remain in good health to effectively get all the nutrition, raw materials, and energy from the food we eat. What you eat and your lifestyle can have a big influence on a healthy digestive system. However, many other things can interfere as well. If the digestive system is injured or compromised, the entire body comes under attack and suffers. This is probably why the immune system is right there on the frontlines intimately involved with the gut.

Gastrointestinal problems among children with neurological conditions

Just how extensive are gastrointestinal disorders in children diagnosed with autism or related conditions? Consider these relatively recent studies. D'Eufemia *et al* (1996) found high intestinal permeability (leaky gut) in 43 percent of autistic children who did not otherwise show clinical gastrointestinal symptoms. If you include those showing other types of intestinal problems, that number would likely be higher.

In 2000, Dr Wakefield a gastroenterologist formerly of the Royal Free Hospital in London, England, and associates published research showing a connection between developmental disorders and gut dysfunction. Looking at 60 children (50 with autism, five with Asperger's Syndrome, two with disintegrative disorder, one with attention deficit hyperactivity disorder (ADHD), one with schizophrenia, and one with dyslexia), they found the majority, as much as 93 percent, had some sort of intestinal pathology.

In April 2002, Uhlmann *et al* published research showing the presence of a pathogen in the guts of 75 (82 percent) of 91 children with both autism-type conditions and having inflammatory bowel disease (an immune-mediated condition). Pathogens were only present in five out of 70 children who did not have a developmental disorder.

In 1999, Dr Horvath *et al* found the following results based on 36 children having an upper gastrointestinal endoscopy with biopsies, intestinal and pancreatic enzyme analyses, and bacterial and fungal cultures. Detailed tests revealed:

- 69.4 percent showed grade I or II reflux esophagitis (25 of the 36 children)
- 42 percent showed chronic gastritis (15 of 36)
- 67 percent showed chronic duodenitis (24 of 36)
- 58.3 percent (21 of 36) had low intestinal carbohydrate digestive enzyme activity, although there was no abnormality found in pancreatic function (19 of the 21 children had diarrhea, 53 percent of the total group)

The authors concluded that unrecognized gastrointestinal disorders may contribute to behavioral problems of nonverbal autism patients.

Buie, Winter, and Kushak (2002) recorded preliminary findings of patients with autism conditions that were brought to them at Harvard University and Massachusettes General Hospital in the U.S. Of 111 patients with complaints of gut pain or diarrhea, they found that most patients had some sort of gastrointestinal problem including

- esophagitis in 23 (20%)
- gastritis in 14 (12%)
- duodenitis in 11 (10%)
- *Helicobacter pylori* bacteria in four
- two with celiac sprue
- five (5%) with eosinophilic inflammation
- ten out of 90 tested (11%) had unusually low enzyme activity; two with total pancreatic insufficiency; five with multiple enzyme defects
- 55% of the children tested had lactase deficiency, and 15% had combined deficiency of disaccharidase enzymes
- 11 of 89 patients (12%) had colitis

Their conclusions were that more than 50% of children with autism conditions appear to have gastrointestinal symptoms, food allergies, and maldigestion or malabsorption issues of some sort. And further large, evidence-based studies are needed in order to fully understand the gut-brain association specifically with these conditions.

Note that these numbers do not reflect all people with autism or related conditions but rather those brought in because of symptoms. So it may just represent a certain subgroup. What is really needed are comparisons to suitable control groups (how do these numbers compare with the general population, other groups, or other children). However, it does reflect that gut problems are common enough and should not be overlooked as a possibility. Neither should unusal behaviors or symptoms be seen as simply a characteristic of autism or behavior problems and not investigated as having a medical basis.

The Pfeiffer Treatment Center in Illinois U.S. reported their clinic evaluated and had records on 503 children with autism. Eighty-five percent exhibited severely elevated copper to zinc ratios in the blood compared to a population of healthy controls (Walsh 2001). Brenner (1979) similarly found elevated copper levels and low zinc in a subgroup of children with hyperactivity.

Many other studies verify that malabsorption and nutrient deficiencies are very commonly associated with neurological conditions (e.g. Nagakura 2000; Oaten 1993). Beshgetoor and Hambidge (1998) note several studies showing increases in copper concentrations in response to stress, inflammation, and infection; in Parkinson disease and diabetes; and malabsorption leading to nutrient deficiencies.

So, it should follow that therapies designed to heal the gut, improve nutritional absorption as well as strengthen the immune system would be extremely beneficial. Digestive enzymes are just such a therapy, facilitating better health on a multitude of levels.

CHAPTER 4

Food and Its Effects on Neurology, the Brain, and Behavior

In pursuing sensory integration therapy, I ran across a particular type of elimination diet. The trick, I learned, was to stop not just dairy but everything that had even a trace amount of the milk protein casein in it, including substances called whey and caseinates. We had tried eliminating dairy previously for Jordan to see if he was lactose intolerant but had not seen any improvement. For lactose intolerance, you need to avoid the milk sugar lactose, but here it said to avoid the milk protein casein. Also, I ran across a lot of information relating chronic ear infections to milk intolerance. Hmmmmm, Matthew lived on antibiotics as a toddler because of a series of ear infections, one right after another. Milk sensitivities also appeared to have a relationship to sensory problems and migraines.

I called the senior occupational therapist at a local center that does the evaluations for children in our town and asked about this diet in regards to sensory integration. She said that removing dairy seems to help some people with sensory issues and it was well worth a try. She did not comment on the gluten. I immediately stopped giving anything with the protein casein.

This was in March 2001. Matthew had just turned nine. After three days, he calmed down, he became more alert, helpful and downright pleasant. The migraines lessened for all of us and the wealth of sensory difficulties improved some more. Jordan, however, still seemed to have significant gut problems. He did, though, volunteer to see if he might be vegetable intolerant and offered to eliminate those from his diet for as long as necessary!

So at least we identified that dairy was a problem, eliminated that, and worked diligently on the gluten-free part. I was not completely convinced gluten was to blame but since the problematic portion of the casein and gluten seemed so similar, it might help to eliminate it too. To be gluten-free, we were supposed to eliminate everything that had a trace of wheat, oats, barley, or rye. Wheat is a possible trigger for migraines.

In addition, I started reducing all artificial chemical additives as well, since these chemicals can be very hard on an already burdened nervous system. Eliminating chemicals such as artificial colorings, flavorings, and preservatives may reduce the total toxin stress load. Several common additives are known to cause migraines and behavior difficulties. Perhaps there was a trigger among them.

Avoiding all gluten was like learning a different language. We tried a vast array of stuff from the health food stores, each tagged with pricing numbers that sent me into sticker shock. I mixed special flours. I tried many recipes. But we were not happy campers. It seemed like most everything tasted like cardboard or pressed sand that dissolved in your mouth, often with a strange aftertaste.

I bought what was supposed to be a casein-free, gluten-free, artificial chemical-free chocolate chip cookie in an attempt to show the boys this wouldn't be so bad. We all took a bite and...'horrid' would be putting it nicely. What was in this stuff, anyway? I looked on the label. Expeller pressed prune puree was the main ingredient. What was *that*?! Followed by banana puree, rice flour and carob. No eggs, no butter, no 'regular' flour, no chocolate. What was being passed off as a good ol' chocolate chip cookie?! My expectation of what was being called food was severely challenged.

The boys were good sports and we eventually found a few things we could tolerate, but it wasn't great. The dairy substitutes weren't such a thrill either. It wasn't that the boys craved a giant glass of milk, but casein and gluten occur in trace amounts in so many foods (not to mention the contamination issues) that you eliminate most everything sold in the grocery stores and restaurants. Now I was really getting a headache!

I also discovered that most people diagnosed with autism have sensory integration issues, although you can have sensory issues without qualifying for something else. I wondered if some of those with other conditions might be suffering migraines internally, and that was behind many of the sensory problems. It is a well-known fact that people suffering migraines are excruciatingly sensitive to light, sound, temperature, and other things. Some seemingly unrelated conditions appear to have some very similar underlying features.

Why eliminate foods – The science behind it

Food and drink, and every little molecule they contain, have an effect on our bodies. Some may be quite meaningless while others have a tremendous effect. Most parents know that what they feed their children affects them, one way or the other, just as you know the difference between eating a bowl of ice cream and a plate of broccoli. If you are allergic or sensitive to some foods, or components of those foods, you may get a bigger reaction from them than you would like.

This response is not imaginary either. Scientific evidence shows that food and chemicals have a direct impact on our mood, our neurological state, our brain, and our behavior. Caffeine and serotonin are a couple of common chemicals that help you feel 'energized' or 'calm,' respectively. Food and behavior is a well-documented relationship. The reference sections in this book list over 100 scientific studies as well as several book titles describing food intolerances and allergies.

Many studies also show the pronounced connections among food additives such as food dyes, nitrites, benzoates, preservatives, and other chemicals to a mountain of conditions, including skin problems, rhinitis, asthma, migraine, severe psychiatric problems, abdominal problems, gastritis, hyperactivity, sleep problems, and mood in children.

Sometimes it takes science a little longer to 'prove' what many know to be true. Halsted (1999) discusses that present-day practitioners of medicine are typically untrained in the relationship of diet to health and disease despite the prevalence of nutritional disorders in clinical medicine and the increasing scientific evidence on the significance of dietary modification to disease prevention. Paying attention to the basic physiology of the digestive tract with the enteric nervous system, a host of neurotransmitters, and the largest part of the immune system located right there goes a looooong way.

Adjusting the diet may improve various conditions and symptoms. The Feingold program helps many with hyperactivity, asthma, and other issues by eliminating various artificial ingredients as well as natural salicylates. Restricting these items benefitted many people before scientific studies discovered a possible biochemical connection as to why. Yeast control diets relieve some neurological effects of *Candida* yeast overgrowth. A gluten-free diet helps those with celiac.

When food adversely affects the brain causing negative symptoms and behaviors, some people refer to this as a cerebral allergy. This means that an allergic reaction occurs in the brain, or affects brain function instead of somewhere else in the body. Cerebral allergies can result in seriously disrupted brain chemistry, moods, and behaviors.

We know migraines can cause major disturbances in the brain with inflammation and swelling, gross alterations in light and sound tolerances, pain, and the ability of the person to function. Would we consider any of these to be sensory dysfunctions? Would head-banging or rocking help? Is there irritability? It may be very difficult for a young child to communicate or deal with such symptoms. Any chemical, pollens, or other substances from the environment can provoke a cerebral allergy reaction as well as foods.

Food intolerances and sensitivities versus allergies

Without going into an in-depth study of the immune system, I will try to summarize the difference in two basic types of negative responses to foods and chemicals. The first is what is called a 'true' allergy. When you are exposed to something you are 'truly' allergic to, your immune system produces certain substances to fight it off called IgE antibodies. An allergy test measures the level of IgE antibodies produced. When a reaction involves IgE antibodies, we refer to that as an IgE-mediated reaction. When IgE antibodies go into action, they usually release a substance called histamine, among other substances. Histamine reactions commonly result in well-known symptoms such as hives, rashes, sneezing, breathing difficulties, mucus production, and other unpleasant manifestations. These are often easier to identify. These are effects that Matthew did not display and so food allergies were dismissed (which may have been correct in the literal technical interpretation but was not as helpful on a practical level).

The other type of reaction involves a different substance known as IgG antibodies. When a reaction involves IgG antibodies, we refer to that as an IgG-mediated reaction. Far more IgG antibodies roam inside our systems than IgE antibodies. About 75 percent of the total amount of antibodies present are IgG, and only a tiny fraction are IgE, and the remainder consist of other types of antibodies.

IgG antibodies are the ones most commonly called up when unidentified particles or chemicals leak through the gut and continue wandering around in the bloodstream. The immune system reacts to these unknown particles by sending in IgG antibodies to deal with them. IgG antibody reactions can be just as serious as the histamine reactions, although maybe not as obviously connected to a particular food or chemical. Symptoms of IgG reactions usually show up as changes in behavior rather than histamine reactions. People refer to these as food intolerances or sensitivities instead of true allergies.

Food intolerances may have either a delayed reaction or an immediate one. You may eat a food today, but not see the reaction until tomorrow, or the next day. Substances can also have a cumulative effect. You may eat a food on Monday, Thursday, and Saturday and be fine. But if you eat this same food in the same quantity on Monday, Tuesday, and Wednesday, you will feel dreadful on Thursday and Friday. This is because you are adding the food to your body faster than the body can process it out. As long as your body processes out the food or chemicals quicker than it builds up, there is no reaction.

This 'tolerance' level will vary from person to person and from food to food. A person's tolerance may even vary depending on other things, such as if the liver is functioning well or if you are sick with a cold. This also contributes to the reason food allergy tests are not 100 percent reliable, or do not definitively let you know if food or chemicals are contributing to symptoms. This does not mean that a particular lab is not doing the test right. It is just that the most accurate tests currently available cannot detect and precisely measure all these possible things going on in the body. There is no hard-and-fast rule that says all IgE reactions result in histamine symptoms, or all IgG reactions are food related. These are the general categories, but overlap occurs from person to person.

To complicate the situation further, a person may react negatively to an entire food or chemical 'family' as opposed to just a particular food. So when I eliminated red food dye and did not see any improvement over one week, it may be because my son was still consuming yellow food dye, or artificial flavorings at the same time. I would need to eliminate all the chemicals in the group and then judge the results. Note: The Feingold program is a short term test you can do to determine if one is sensitive to certain artificial chemicals and naturally occuring salicylates. (see www.feingold.org)

Enzymes are particularly effective on IgG-mediated reactions because these reactions are often based on insufficiently broken-down food. Enzymes break these food particles down directly and thus eliminate the negative IgG reaction (Oelgoetz *et al* 1935; McCann 1995). Enzymes are not generally as effective on IgE-mediated reactions because these do not always relate directly to food particle size. These are general guidelines and will vary by person. Enzymes may lessen the effects of IgE reactions because of their support for the immune system and in decreasing inflammation – a more indirect way of reducing symptoms. True allergies can be very serious so great caution is advised with these.

Casein, gluten, and other proteins

Casein is a protein in dairy products (different than lactose, the milk sugar). Gluten is a protein from small cereal grains such as wheat, barley, rye, and oats. Other foods or sources may also produce similar protein structures, such as corn, soy, yeast, or bacteria. At times, the body itself produces these structures internally. Proteins are broken down into smaller chunks called peptides, or directly into amino acids. Peptides are then broken down further into individual amino acids as well. The whole proteins and the completely broken-down amino acids are not considered potential problems, only the intermediate partially broken-down forms. These intermediate peptides may be of different lengths and shapes. Some peptides may inappropriately pass through the gut lining and trigger an IgG or other type of immune system reaction, just as other foods do.

One highly contested theory speculates that when gluten and casein proteins are not completely broken down and get absorbed into the body, they might travel through the bloodstream and affect the brain. This theory suggests one of the possible shapes a peptide may end up with is a structure that fits into receptors in the brain that may trigger certain moods and behaviors. These particular receptors are referred to as opiate receptors because they also interact with other opioids like morphine. However, substances classified as opioids or having opioid properties are different. Opioids from foods do not have the same properties or cause identical reactions as morphine.

There are natural beneficial effects of these opioids from food. Some opioids are needed so the body operates properly and the nervous system functions well. Some opioids facilitate proper digestion. Opioids have a known modulating effect on digestion, specifically with amylase enzymes and bicarbonate activity. But exactly how this plays out in neurological conditions, and particularly in autism, is still under investigation (Dooley, Saad, and Valenzuela 1988).

We need to consider if the opioids are affecting neurotransmitters and/or receptors in the gut, in the brain, or both. When is the effect beneficial and when is it unwanted? Can levels be 'adjusted' without causing other problems? Neurotransmitters need to be in balance. With neurotransmitters located in both the gut and the brain and interconnected to one another, it can be hard to isolate exactly what is going on and how best to manage it.

Researchers have spent many years looking into the possibility of excess opioid peptides with conflicting and controversial results. Research also implicates possible peptide interaction with receptors in adult schizophrenia, phenylketonuria, postpartum psychosis, and some with celiac disease who consume gluten (e.g. Gobbi *et al* 1992; Hallert *et al* 1982; Seim and Reichelt 1995; Hadjivassiliou 2001).

Some groups refer to these food-related effects as allergy-induced autism, or a cerebral allergy response. These cases could be a food allergy or intolerance, which is expressed differently in individuals and leads to a wide range of reactions, both physical and behavioral.

Parents see different responses as well. Some say their children seem better without gluten or casein foods (although the reason is not known), some say their children are much better when consuming gluten and casein foods, and some say it doesn't make any difference whether the foods are given or not.

Several studies looking specifically at this issue support finding no opiate problem (e.g. Hunter *et al* 2003). Others indentify several types of casein, of which only a few may be a potential problem. Care must be taken to correctly correlate cause and effect, and not just jump to one conclusion. You might see improvements from food removals for a number of reasons. Dairy is known to trigger a histamine response in some individuals. Or you might have lactose intolerance. There might be artificial additives in the dairy product selected that are the problem and not the dairy itself. Also, there are so many neurotransmitters in the gut, it is not always easy to separate what is speculated to be a food reaction from something affecting one of these neurotransmitters.

One way to reduce any suspected problematic peptides may be by eliminating casein and gluten containing foods from the diet completely (or other suspected foods). Undertaking such a casein-free, gluten-free diet has been beneficial for some with autism conditions, reducing a number of the seen problematic behaviors and concerns (Cade 2000; Lucarelli 1995). But why removing dairy, grains, or other foods may be beneficial to some is unknown and needs further investigation.

The disadvantage of food eliminations is that it is often incredibly difficult to completely eliminate all the dairy and grain sources, as I quickly discovered. If other foods are creating these peptides, like soy or corn, you will need to totally eliminate these foods as well, if you can figure out what they are. However, food removal will not have any effect on peptides generated within the body.

When you start eliminating foods, especially entire food groups, you are also eliminating all the nutrition they contain and energy they supply. Such a restrictive diet can be a far more rigorous and difficult undertaking than it may seem on the surface. One that is often financially expensive to do as well.

Another way to accomplish the elimination of insufficiently broken-down peptides is to take digestive enzymes specifically formulated for breaking down these proteins all the way beyond the intermediate stage. Enzymes break down the peptide chains so they do not exist in a form that may potentially cause a problem. So enzymes may help no matter which reaction is the suspected problem, or where in the body it happens. Enzymes help release the individual amino acids from peptides so the body can use them for the raw materials it needs. Several amino acids are pre-cursors for important neurotransmitters like norepinephrine, serotonin, and dopamine. An amino acid deficiency may be helped in some too.

Another advantage of digestive enzymes is that by taking the appropriate ones, they will break down all foods and not just the ones you think may be causing a reaction. They can digest peptides created by external sources as well as internal sources. Enzymes are also easier to fit into regular daily activities rather than attempting to work around every possible source of casein, gluten, and other foods.

The search for a digestive enzyme product to break down the intermediate peptides produced by gluten, casein, and other possible proteins started as soon as these peptides were first suspected as possible causes of some of the undesirable behaviors. However, finding someone who could actually make such a product turned out to be an unexpectedly difficult task. After a great deal of effort, several years and a worldwide search for just the right scientist, the product SerenAid came to market in 1999. A few other products followed. However, none were adequate to the extent that someone could avoid food eliminations. In April 2001, Peptizyde came out (formulated by the same scientist that came up with SerenAid). Peptizyde has been the breakthrough blend effective enough to allow many people (although not everyone) to consume casein and gluten regularly with these enzymes, should that be their choice.

Another course of action to prevent insufficiently digested foods from escaping into the bloodstream is to improve gastrointestinal health. In particular, healing a damaged or leaky gut would prevent *any* inappropriate substances from leaving the gut in the first place.

Determining food sensitivities

A process called an elimination diet is a very reliable method of finding out if you react negatively to a particular food. This involves consuming only a few foods for at least a week or so to see if symptoms improve, and then adding back in one food at a time to ensure it is tolerated. Remember that some food reactions are cumulative or delayed over a few days. You may need to remove a food entirely for a week, and then reintroduce it for a few days in order to see if there is a reaction. Then you need to watch for entire food families and chemical groups. Some standard allergy tests are available, but these are not always conclusive or definitive. If you run any tests, or are considering any medications, check to see if any supplements you are taking may influence the results.

Common problematic foods tend to be whatever is eaten the most. Among people on a typical American diet (US), these foods include milk, wheat, eggs, soy, and sugar. In Japan, more people are allergic to or intolerant of rice, and in Scandinavia it is fish. Some babies that start out on cow's milk become intolerant of it; however, among babies started out on soy milk, there is a corresponding increase in soy allergy and intolerance.

A rotation diet is another method for reducing food intolerances from building up, or minimizing reactions. This program stresses only eating one food, food group, or chemical group every four days or so. An example would be to have soy milk and wheat flour on Monday and Thursday, cow's milk and sorghum flour on Tuesday and Friday, fruit juice and potato flour on Wednesday and Saturday, and rice milk and tapioca flour on Sundays. The idea is to keep rotating foods so no one food type is eaten excessively.

There are different versions of rotation and elimination diets. Most books and specialists dealing with nutrition or allergies have sample programs or guidelines. When greatly changing a familiar diet, withdrawal symptoms, adjustments, or different behaviors may appear. These may include moodiness, aggression, headaches, solitude, sleepnessness, tantrums, and crying. Reducing the food sources at a slower rate may be helpful in easing these reactions.

CHAPTER 5

Intestinal Permeability
– The Food-free Diet

My husband, our neurologist, family doctor, and I were all very, very concerned about removing entire food groups from our diet - particularly for growing kids. Food tends to work as a whole. Giving individual supplements does not necessarily produce the same benefits, that is, if you can even adequately identify what it is you need to replace. I was reading all I could from every source and especially about other parents' experiences on the Internet. What really alarmed me was hearing how parents would remove all casein and gluten foods, only to find out they needed to remove all soy...then all corn...then all highly phenolic foods...then all yeast...then sugars...etc., just to keep at the same level of improvement. It seemed like a never-ending progression of removing foods upon foods.

And then, because you were eliminating so many nutrients you needed to buy and add supplement upon supplement to replace the vitamins, minerals, and everything else you had just removed. This 'diet' was taking on a life of its own...and the life it was taking was mine! Our whole family was disrupted! It was like this huge mammoth had moved in, and everything we did revolved around it. This did not make much sense to me and was a dismal undertaking.

The most common logic would be, 'If a certain food makes me

ill, I should not eat it.' This may work in some cases, however this strategy falls apart in other situations. If a person has a damaged intestine or 'leaky gut,' then insufficiently broken-down food particles and other things may be crossing into the bloodstream when they shouldn't. Then the immune system reacts. This is how a leaky gut very commonly leads to all types of food and chemical sensitivities.

Usually you will see negative reactions to whatever foods a person eats the most. Eliminating these foods may bring about some relief, and so you say the individual is intolerant, sensitive, or allergic to those foods. You start eating more of other 'safe' foods. But since the gut is still damaged and 'leaking,' now those foods are insufficiently digested, get into the body, and start causing negative reactions. Even if you now eliminate those previously safe foods, you will very likely end up with a problem with the next food you substitute in. Eventually, devoted parents are up late at night on the Internet trying to find a reliable source for free-range buffalo and cassava root flour.

Trying to balance nutritional needs is particularly a problem with children because of the limited types of food they tend to accept. My children are pretty exceptional (of course!), but if they need to get more sulfur in their system, they will hardly accept creamed Brussels sprouts as a good substitute for scrambled eggs.

A word of caution here: there is a strategy of trying to hold out on food with the idea 'the child will not starve' and eventually will eat the substitutes you offer. This may not always be true. I did this once with each of my boys, and around 24 hours later hunger got the best of them and they ate the food offered. Afterwards, I freely offered my insightful tactic to others demonstrating how I had wisely nipped that in the bud, remembering that 'the child will not starve.'

Sometime later I followed the story about a mom who ended up taking her child to the hospital because he would not eat the new foods. I thought that child must have some special problem. But the next time this came up, I took notice. By the third, fourth, and fifth times a child was hospital bound because they would not accept soy milk or other foods, I realized how short-sighted that line of 'the child will not starve' was in these cases. The tyke may be in real trouble. Those who do not have fully developed, properly working neuro-feedback mechanisms may not respond the same as the majority of people who do – those 'typical' ones for whom sweeping, blanket statements may be based on. So if you feel your child is not responding appropriately when holding out on food, please consider that he or she may legitimately not have an adequately functioning response system.

Many supplements need to be dosed throughout the day. I got to the point of feeling rather silly taking out food after food and then giving handfuls of vitamin capsules. It just didn't make sense. And it turns out there was a good reason it did not make sense. In many cases, the fundamental problem may not be that the person truly has 92 basic foods that he will never be able to eat. The root cause may be that his gut is injured which results in food intolerance and other problems.

Leaky gut – The science behind it

The conditions of leaky gut, food intolerances, and yeast overgrowth are intricately connected. Any one of these may lead to the others, and these conditions build upon one another. Because these three issues affect the gut, they also have a profound impact on the nerves, the brain, and behavior. A study by D'Eufemia *et al* (1996) looked at intestinal damage in 21 children with autism conditions using the intestinal permeability test as an indicator. They found that 43 percent of the children had damaged intestinal linings, but there were none in the 40 control individuals. Although the number of children in the study was not high, it indicates that a greater probability of leaky gut may exist with these conditions than in the general population.

In a 1999 review of the scientific literature on intestinal permeability, leaky gut, and intestinal disorders, Hollander found ample evidence indicating that most patients with Crohn's disease have an increase in intestinal lining permeability 10 percent to 20 percent greater than in their clinically healthy relatives. Another review found intestinal permeability measured conservatively to be around 74 percent or greater in people with untreated or silent celiac disease (Vogelsang, Schwarzenhofer, and Oberhuber 1998). Research identified leaky gut as a factor in celiac disease 20 years ago as well as with irritable bowel syndrome.

Even much of the research speculating on the excess-opiate peptide theory actually finds the basic problem to be a leaky gut and insufficient digestion, and not an inherent problem with casein and gluten containing foods (Whiteley and Shattock 2002; Reichelt 2003).

The intestines are permeable to a certain degree so the proper nutrients get through. Spots in the gut wall open and close selectively to admit the good stuff appropriately as the contents of the intestines pass by. Normally, only certain nutrients are absorbed if they are sufficiently broken-down and in the right form. Everything else is selectively blocked out. But when the pores are too big or the screening process breaks down, the intestines become hyperpermeable (overly permeable). Leaky gut syndrome is a term used when the intestines become damaged, more openings develop in the gut wall, and the wall becomes more 'porous' to the extent that some of the contents passing through the intestines are allowed to get into the bloodstream when they should be kept out. The contents of the gut are 'leaking' through. (Hemmings and Williams 1978; Loehry et al 1970; Warshaw et al 1974)

And not just food particles slip through. Pathogens, toxins, and other types of 'waste' get through that should normally be screened out (e.g. Wolf et al 1981; Walker 1975). Insufficiently broken-down food particles or toxins may cause the liver to work much harder trying to clean everything out. The liver is the headquarters of chemical reactions in the body. It performs a multitude of tasks, including cleaning out toxins, manufacturing compounds for the body out of raw materials, and breaking down other substances into usable materials. The liver may not be able to keep up with all the detoxification demands sent its way.

A sluggish, overworked liver can lead to all types of problems. Just imagine if the giant New York Grand Central Train Station got all backed up. One clog in the system leads to a delay farther down the track. Then the other systems are off schedule and cannot deliver their raw materials on time. The body cannot manufacture essential materials on schedule. The toxins start building up in the body. The blood needs more cleaning, and nutrients back up just waiting to be processed so they can continue their journey to other parts of the body. The biological pathways waiting for nutrients come to a halt. Normal biological functions break down.

When the gut becomes hyperpermeable, all sorts of gunk can get through and run loose in the body. When this happens, the immune system kicks into gear to stop these invaders. Agents of the immune system latch on to the invaders – like a policeman handcuffing himself to a crook. This bonding is called an immune complex.

Typically the immune complex, or the unattached invading particles, should be escorted out of the body quickly. However, this task often falls to the liver, which may be quite overworked and not able to do this immediately. If not removed in a timely fashion, the troublesome particles and complexes can migrate through the body and settle in any of the different tissues they pass by. This leads to inflammation in whatever part of the body they settle.

Now we have a new problem: inflammation. This puts even more pressure on the immune system to cover even more ground in defending the body. With the immune system running on 'high' on a regular basis, it may be spread thin over a wide array of territory defending the gut, cleaning the blood, fighting inflammation, warding off pathogens, and so on. Many autoimmune conditions start this way. Which type of autoimmune condition a person ends up with may depend on which part of the body the immune complexes settle into.

Problems of having leaky gut

So a few food particles slip into the bloodstream, is that so bad? Well, frankly, yes it can be very bad. Leaky gut is a condition that can directly lead to many other specific disease states, or indirectly aggravate or worsen other conditions (Zanjanian 1976). You may have leaky gut and not be aware of it. Many food intolerances and sensitivities are a consequence of a leaky gut to some degree. The invaders might cause 'brain fog' or hyperactivity. You may just feel run down, out of energy, have many food and chemical intolerances, or a multitude of other seemingly unrelated problems. Symptoms of leaky gut syndrome may include:

- aggression
- anxiety, nervousness
- asthma
- atypical sensory reactions
- bladder infections, bed-wetting
- bloating or gas
- chronic joint, muscle, or abdominal pain
- confusion, memory problems
- diarrhea or constipation
- fatigue, poor exercise tolerance
- fevers of unknown origin
- fuzzy thinking or 'brain fog'
- indigestion
- migraines
- mood swings
- poor immunity
- skin rashes

As the harmful material continues to leak through the intestines, the consequences become more severe. Some of the conditions that may be connected to or aggravated by intestinal permeability follow. Many are classified as autoimmune conditions.

- autism/Asperger's Syndrome/PDD
- autoimmune conditions
- celiac disease
- chemical sensitivities
- chronic fatigue syndrome
- Crohn's disease
- cystic fibrosis
- digestive disorders
- dysbiosis
- environmental sensitivities, allergies
- food sensitivities, allergies, intolerances
- hyperactivity (in general and AD(H)D)
- inflammatory joint disease/arthritis
- intestinal infections
- irritable bowel syndrome
- liver problems

- malnutrition
- pancreatic insufficiency
- schizophrenia
- skin conditions
- ulcerative colitis

And remember that some important enzymes are produced and work at the point of the mucosal gut lining (for particular starches, sugars, small peptides, and fats). A compromised or injured intestinal lining means certain enzymes will not be available and digestion is impaired.

Malabsorption and nutrient deficiencies

The main sources of malabsorption are identified as maldigestion, bacterial contamination, and mucosal abnormalities (Salerno *et al* 2002). Besides particles being too large, escaping into the bloodstream, wandering loose in the body, and causing havoc, there is another side to consider. Because the food was not digested and absorbed properly, the person may experience nutrient deficiencies. In one area, the nutrients are bound up in a manner the body does not recognize and is hailing the immune system to remove them. Yet, in another location, the body is starving for those same nutrients. The biological system is overworking unnecessarily in one area while shutting down somewhere else because of insufficient raw materials.

Vitamin therapies may target some of these other systems starving for raw materials. By supplying just the one or two specific raw materials, we may temporarily fix that part, if the vitamin can even reach the intended area of the body, but the system as a whole is still malfunctioning. If the gut is not absorbing food well, then the supplements may not be absorbed well either.

What causes leaky gut?

Several factors can lead to leaky gut, either occurring individually or working together. Any chemical or physical activity that stimulates the pores in the intestines and keeps them open for too long can lead to increased permeability. Some common sources follow.

Yeast – Yeast or bacteria are main causes of leaky gut. Yeasts are single-celled are single-celled organisms that usually reside in the mucosal lining. A healthy immune system and beneficial bacteria keep them in balance. However, when the immune system becomes weakened or is focusing on other invaders, yeast can grow out of control. Yeast can grow out of the single-cell form and into a fungal form. The fungal form grows root-like tentacles (hyphae) that drill deep into the mucosal lining, poking 'holes' in the gut.

Dysbiosis – Dysbiosis is when the internal gut microbial populations are not in sufficient balance for good health. Besides the physical damage created by an imbalance of intestinal microbes, bacteria and yeast give off many toxins. (See Chapter 14 on Dysbiosis)

Environment toxins –With the body's detoxification system overloaded or dysfunctional, environment toxins from either inside or outside the body may build up. Some of these compounds may irritate the intestinal lining. The constant inflammation and immune system activity produced can create a more permeable gut wall.

Chronic stress – Stress suppresses the immune system and can alter intestinal physiological function, increase gut permeability, and cause inflammation (Hart and Kamm 2002). A healthy immune system can easily block out typical pathogens, but a weak one may be overrun.

Inflammation – Anything causing inflammation may lead to leaky gut. This can result from insufficiently broken-down food or infections of any kind. Maybe the stomach just does not produce enough hydrochloric acid, resulting in improper digestion. Larger particle sizes from any food can irritate the gut lining. Another cause of inflammation in the intestines can be bacteria, yeast, or other pathogens.

Constantly active immune system – When the immune system is overactive for an extended period of time, leaky gut can develop. If you have pathogens and toxins in the gut, the immune system will be sending in the appropriate agents to bind the toxins up (immune complexes). If too many immune complexes and agents are constantly floating around the mucosal lining, this may cause inflammation. The inflammation can get worse when the gut lining does not have a break to rest and heal. Thus, the lining wears thin and loses integrity allowing particles, chemicals, and debris through. Most people eat several times a day. This means the immune system must be on constant alert picking up all the gunk and foreign material entering the blood system. Even if the immune system is doing a stellar job, to maintain all the effort expended in protecting the body requires a lot of nutrients and energy from the rest of the body that could be going elsewhere.

Medications — Medications including over-the-counter pain relievers like acetaminophen or ibuprofen may lead to increased permeability. They are considered 'hard' on the gastrointestinal lining. Aspirin reduces the thickness of the mucosal lining in the gut, thus making it more susceptible to yeast, inflammation, and irritation by food passing through. Some prescription medications can also stress the gut.

Diet — A diet high in sugar, refined flour, and processed foods, complete with their rainbow of artificial food colorings, flavorings, and preservatives places a lot of stress on the immune system as well as the liver. Our bodies see most of the artificial chemicals as pure toxins. The more you consume, the more the body must process out. Most processed foods have less nutrition in them as well. So, you use more energy and nutrients processing these foods from your body than you receive from them. A diet high in refined sugar, flours, and artificial chemicals can actually cost you nutrients and energy rather than supply them. In addition, the pathway needed to process these toxins may be dysfunctional in many with neurological conditions (Hollander 1999). (see Chapter 17 Sulfur, Epsom salts, and Phenols)

Zinc deficiency — An interesting finding is that many individuals with autism conditions seem to have a zinc deficiency. Zinc is necessary in maintaining intestinal wall integrity. Supplementing with zinc could contribute significantly to healing a leaky gut in about eight weeks (Sturniolo 2001). Zinc is also instrumental in a maintaining a healthy immune system (Prasad 2002). The synthesis of serotonin involves zinc. Since serotonin is also necessary for melatonin synthesis, a zinc deficiency may result in low levels of both of these compounds, which are essential to the sleep cycle and calming.

Who may have leaky gut?

As you can detect from the above list of aggravating causes, practically anyone can develop leaky gut. In fact, it is becoming more and more common in the general population just because of our lifestyle and diet. People with weakened immune systems, compromised nerve systems, or faulty digestive systems may be more susceptible.

The Intestinal Permeability Test is a test that can give you an idea of how permeable your gut is. This is available at some commercial labs, or you can check with local labs and doctors to see if they offer it. If you have a yeast or bacteria overgrowth, you have a very high probability of also having leaky gut, and most likely an assortment of food intolerances as well. The person who reacts to 'everything,' or takes allergy tests that come back saying they are reactive to 87 different substances, probably has a leaky gut to some degree. Many of those allergenic responses may be due to insufficient digestion and stuff passing through which is usually blocked out by a healthy intact mucosal lining. Many people report that once their gut heals, they experience an amazing recovery from allergies, food intolerances, and multiple chemical sensitivities.

Fixing the hole in the boat

If you have a boat that leaks, you can bail and bail. You can even buy a bigger bucket for bailing. But until you fix the hole in the boat, you will be bailing forever and uncomfortably wet. Some of the solutions people pursue just address the superficial symptoms. Often better and quicker results will be seen if healing the leaky gut condition is part of the overall treatment program, rather than just focusing on treating the various symptoms that result from the injured gut. You may add some supplements that compensate for some of the damage created by nutrients not delivered on time, and even see some positive results doing this. However, if you do not fix the core problems, the leaky gut will continually trouble you. By focusing on healing the gut instead of just removing foods or adding in tons of supplements, you may see an end to all those food intolerances and the behavioral, learning, and health problems that result from them.

Constant supplementation seems to be a common practice with treatment for the autism spectrum and other neurological conditions. You have a child who is obviously ill and exhibits a number of behavioral symptoms that you would like to stop. Maybe you hear that supplementing with a particular nutrient will help a particular symptom. Say, Nutrient E helps maintain eye contact. So, you rush out, buy some Nutrient E, and start supplementing the diet with it. You may even see some improvement in eye contact. Then you hear about another supplement, Nutrient H, that helps a different symptom, say hyperactivity. So, you go get that. This continues with supplement after supplement trying to fix symptom after symptom, until you are giving handfuls of supplements that are only trimming away at the symptoms, not fixing the problem at the root level. This is like watering a thirsty plant leaf by leaf.

Your boat may be staying afloat as you bail with different buckets. Or you may just have slowed down the sinking process, and convinced yourself that a life of treading water really isn't all that bad. You need to have a better plan to fix that hole in the boat. Digestive enzymes may very well be a better plan. In some cases, they can contribute to fixing that hole directly, such as breaking down food or reducing inflammation. In other cases, enzymes may just be a better patch until other therapies come along that finish the job, such as aiding digestion until you can remove any toxic materials or control a yeast overgrowth. At the very least, enzymes may be a much better bucket – like watering the plant at the roots.

Digestive enzymes may be extremely helpful with leaky gut situations because they tackle the problem on several fronts. Enzymes break down the food particles so they do not exist as larger particles that will physically irritate the gut lining or activate the immune system. Plant or microbe (fungal)-based enzymes are especially effective because they do much of this breakdown in the stomach before the food even enters the intestines, a good 60 to 90 minutes before pancreatic enzymes emerge on the scene. Then, by breaking the food down, enzymes are also freeing the individual vitamins, minerals, and other nutrients so the body can use them as the raw materials it needs as well as releasing the energy from your food to you. Normal biological processes can proceed. Because your nutrition comes from food, you then do not have to supplement extra vitamins, minerals, and whatever else to make up for what you eliminated through diet. Nutrition from whole foods is generally more effective than from many supplements too.

Because enzymes can process the food particles down to their essential forms, anything that does leak through the gut while it is healing is less likely to provoke a negative reaction. Enzymes work on foods you do suspect as well as those you don't, or from unknown sources. Food intolerances usually drop off dramatically when enzyme use begins. Many people report improvements the very first day of taking enzymes. However, some food intolerances are processed out of the body a few days later, so it is also common to hear of significant improvement by the end of the first week on enzymes.

Next, enzymes proactively support intestinal health. They can act as trash collectors removing dead tissue, debris, chemicals, and toxins from the body. This cleaning out allows the gut to heal faster. Another bonus is that enzymes are effective at clearing out pathogens that may cause and contribute to damaging the gastrointestinal tract. Bacteria and parasites are made of proteins, viruses have protein coatings or 'films,' and yeasts have outer shells consisting of cellulose and protein. Proteases and cellulases can help break these intruders down, and then carry off the toxins and dead cells the destroyed pathogens leave behind.

If the digestive enzymes themselves are absorbed into the bloodstream along with the other things, this can be very beneficial. Enzymes, especially the proteases, travel through the bloodstream cleaning out any gunk, toxins, and waste that may be accumulating there. They selectively latch onto toxins and escort them out of the body leaving the good tissue and red blood cells to carry on. This assistance in cleaning the blood helps relieve the burden on the liver and the immune system. Enzymes help clear out the traffic jams so everything can get back on schedule. Proteases are often given in between meals for just this purpose. If you give proteases with food, the enzymes will act on the food first, so giving them between meals sends them directly into the bloodstream to do cleanup. Substantial scientific research has established that the bloodstream takes up enzymes intact where they work in this way (Leibow and Rothman 1975; Rothman, Liebow, and Isenman 2002).

Another benefit of taking proteases between meals is to reduce inflammation. Bromelain and papain (protease enzymes derived from pineapple and papaya) have undergone study in great detail for this purpose and are found to be incredibly effective at reducing inflammation. Proteases can reduce inflammation in the gut directly. Bromelain and papain are well-known agents for assisting in healing gastric ulcers (Monograph: Bromelain 1998).

Material leaking through the intestinal lining can make its way to joints and tissues and aggravate them to the point of inflammation, or add to any inflammation already started. Immune complexes can settle or be deposited in joints or tissue as well. The proteases in the bloodstream break down these sources of inflammation as they pass by in the bloodstream. Then the debris is transported out of the body, freeing the immune system up to do other things, and allowing you to enjoy your life without so much pain. (Boyne and Medhurt 1967; Seligman 1962; Starley *et al* 1999; Tassman, Zafran, and Zayon 1964; Wimalawansa 1981)

Besides enzymes, beneficial bacteria (probiotics) are also very helpful in healing a leaky gut (Linskens *et al* 2001). A good probiotic will help restore the needed balance among the bacteria, yeast, and other microorganisms. Out with the bad and in with the good. A consistent supply of probiotics is like tending a lawn. You need to kill the weeds, and then keep seeding with the 'good' grasses. A healthy lawn will need just a little maintenance. However, doing no maintenance will very soon give you a yard overrun with weeds again.

Taking enzymes regularly with food and on an empty stomach are primary ways of proactively supporting gut health. The nice thing about enzymes is they address the damaged gut, problematic peptides, food intolerances, and nutrient deficiencies all at the same time.

Because enzymes may be recycled in the digestive tract, you get a good value from your enzyme supplement investment – more than just a one-trip excursion through the digestive tract. Additional zinc and drinking plenty of water to keep things flushed out may also assist in gut healing. I found that the average time for a leaky gut to heal is about three to six months, and going up to 12 to 18 months for more severe cases. The gut is in a constant dynamic state and starts healing immediately.

A study by Lucarelli in 1990 showed that 87 percent of 49 children with migraine headaches and testing positive to one or more foods improved after following an elimination diet for four to six weeks. Thirty-one children recovered completely following the elimination diet and nine improved. The children were able to reintroduce the suspected foods 6 to 12 months after implementing the elimination diet. This would be a sufficient time for intestinal damage to heal and then formerly problematic foods might be added back.

Another interesting find is that oats facilitate gut healing (Keshavarzian *et al* 2001). Thus, those consuming oats with enzymes may have faster gut healing than those that do not. At our house, we do eat a great deal of oats. Oats turned out to be an interesting twist not only in our story but with another special little boy later on.

CHAPTER 6

What are Enzymes and What Do They Do?

By this time, going with digestive enzymes seemed to be a far more practical and efficient alternative than the other choices at hand. When I first started considering enzymes, I researched the issue as I do most things…intensely. (Guess I get it from my son!) I had been developing educational materials in technical and biological subjects for some years, so the topic of enzymes was not new to me. However, looking at them as health supplements for people was.

Enzymes are proteins produced by all living organisms, and, like all proteins, they consist of amino acids. What makes these proteins different from other proteins is how they behave in the body. By definition, enzymes are catalysts that make many essential biochemical reactions 'happen' and are not used up or chemically altered in the process. As a catalyst, they help a chemical reaction take place quickly and efficiently. Some reactions would either happen very slowly or not occur at all without enzymes. So a little bit of enzyme can effect a big change.

The same variety of amino acids that occur in all living things make up enzymes. The amino acids connect in particular sequences to form protein chains. The amino acids in the chain often bond together creating folding patterns and twisting into certain shapes.

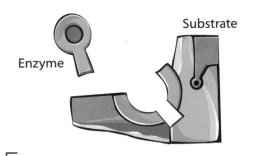

Enzyme

Substrate

Substrate - a protein
consisting of various amino acids

Enzyme connecting to substrate

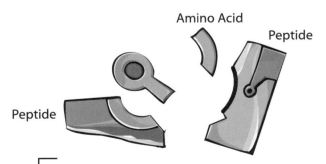

Amino Acid

Peptide

Peptide

Enzyme Breaking Substrate -
amino acids and peptides from protein;
enzyme remains unaltered, continues on

Figure 6.1

Many times the sulfur atom in an amino acid in one part of the chain will bond with a sulfur atom in another part of the chain. This creates a disulfide bond (meaning two sulfurs) that holds the enzymes together. The particular folding pattern of each enzyme gives it distinct characteristics and functions. When anything disrupts the specific folding pattern, the enzyme often loses its ability to function, becoming inactivated or destroyed. Each type of enzyme has a special function and works in a particular way. Enzymes are essential to every aspect of life and carry out all the daily biochemical functions. They are the basic elements that activate all functions in the body, facilitate reactions that build compounds from the body's raw materials, transport elements throughout the body, break down substances, and eliminate many unwanted chemicals in the body.

Enzymes are chemicals that facilitate other chemical reactions. Food itself is essentially just a mixture of chemicals that are broken down by enzymes (see Figure 6.1). The released nutrients are the raw materials. Vitamins and other nutrients cannot work in the body by themselves. They require enzymes to transport them throughout the body and make use of them. Enzymes unlock the benefits of vitamins, minerals, proteins, and hormones and put them to work in the body. Enzymes are the workers and assist many biological, chemical, and metabolic reactions, but are not 'alive' themselves.

Sometimes particular enzymes need certain vitamins and minerals in order to function. Magnesium participates in over 300 enzyme reactions. These additional elements are called co-enzymes. A co-enzyme may give the enzyme the three-dimensional structure it needs to create the 'active site' necessary to perform its catalytic function. If a needed co-enzyme is not available, the enzyme will not function.

The right enzymes for the right food

Enzymes are very specific in their function, one of the few known and predictable things in biology. The substance they act on is called the substrate. One of the fundamental things about digestive enzymes is you need to take the right enzymes for the right food types. Having the right enzyme is like having the right key to open up a locked door. The correct key opens it right up. However, if you use that same key in the wrong door, you will be standing out in the rain all day fiddling and fumbling with it because it just won't work.

The traditional explanation for enzyme action is the Lock and Key (a more refined, newer analogy is the Hand and Glove). Think of an assortment of enzymes as keys on a keyring. Each specific enzyme (key) works in one specific way on one specific substrate (the lock). You may have several enzymes (keys) but only one may work on a particular food (lock). Just as one key can open many doors that all have the same type of lock without destroying it in the process, one enzyme can work many times on the same type of food without being destroyed. If you do not have the right key (enzyme), you can't open the lock no matter how many other keys you have. In the same way, you cannot digest a particular food without the right enzyme for that food no matter how many other types of enzymes you take.

Let's say we have a food with a basic structure as follows where the letters represent various amino acids:

A-A-A-A-B-B-C-C-A-A-A-A-B-B-C-C-
A-A-A-A-B-B-C-C-A-A-A-A-B-B-C-C

Situation 1

Now we have an enzyme that only breaks the bond between the structure B–C (B bonded to C). If we take this enzyme, the food will be broken this way with every bond between a B and C eliminated:

A-A-A-A-B-B   C-C-A-A-A-A-B-B
C-C-A-A-A-A-B-B   C-C-A-A-A-A-B-B   C-C

Situation 2

If we have another enzyme that only breaks duplicate type bonds (it breaks any bond between two similar amino acids), the food will be further broken down and look something like this:

A A A A-B B C C-A A A A-B B
C C-A A A A-B B C C-A A A A-B B C C

This enzyme has a slightly wider range in the types of bonds it will break than the more specific one in Situation 1. It breaks bond between two As, two Bs, two Cs, or any other two duplicates whereas the enzyme in the first situation only broke one type of bond.

Situation 3

This amount of digestion or breaking down of the food may be quite satisfactory for many. But if an individual cannot handle even a little bit of unbroken food particles, he may still have a negative reaction because of the unbroken A-B and C-A bonds. So, if a particular enzyme product only contains the enzymes that result in Situations 1 and 2, our friend eating the food may not show much improvement. He may not even need those two enzymes at all, just something that will break the specific C-A bond. So now our friend needs to look for a product that contains this specific enzyme. This also illustrates why enzymes rarely, if ever, directly interact with other supplements and medications. Each one only works on specific types of bonds.

If a little bit of enzyme can work on a large amount of food, why don't we take one capsule for the day…or even for an entire week? Why would we care which product has more enzyme activity if we only need a small amount? The answer lies in the function of digestion. Food and enzymes are only in the stomach for about 60 to 90 minutes and then they pass into the intestines where most of the absorption takes place. Any insufficiently digested particle entering the small intestine is a candidate to pass through a leaky gut, invoke an allergic reaction, or cause inflammation. We want to ensure that all the food is sufficiently broken down, or as much as possible, before it enters the small intestine and is absorbed into the body.

This is comparable to one person with one master key attempting to unlock all the doors in a large office building. This industrious fellow can do it with the key given an unlimited amount of time to get through the building. If he only has 90 minutes, he may very well only get some of the doors unlocked. If a few of his buddies (more enzymes) came along with additional master keys, then the group of them could get all the doors unlocked in time.

So, getting the right type of enzyme for the correct type of food, and having enough present to digest the quantity of food in the given time are the 'keys' to getting the best results. The Guide to Comparing and Buying Enzymes in this book (Appendix A) presents a list of the different types of enzymes and which foods they break down. The seven steps given outline how to read an enzyme label, calculating costs, and other considerations for evaluating enzyme products.

Explaining enzyme function to children

Kids may cooperate more with taking enzymes and monitoring their food if they know what is going on. I explained how enzymes worked to my boys like this. I took some of their stuffed animals and put them end-to-end in curvy lines. The animals [a] represent the individual amino acids.

          a-a-a-a-a-a-a-a-a-a-a-a-a-a-a-a-a-a-a

Then I said, 'Our bodies need the amino acids to work right. When we eat food, the food looks like this in our stomachs. The food is still too large like this, so our stomachs try to break the food apart to get the individual amino acids.' Then I took my hand like a knife and chopped the lines apart.

          a-a-a-a-a   a-a-a-a-a-a
          a-a-a   a-a-a-a-a

Then I said, 'The enzymes in our stomachs do this. The enzymes chop the amino acids apart. If the lines are not chopped apart, we get sick and feel bad.' Do more chopping on two lines – the more dramatic with lots of sound effects the better.

          a-a-a-a   a   a-a-a-a   a-a
          a-a   a   a-a-a   a-a

Then I said, 'But sometimes our bodies do not have enough enzymes to do all the chopping so we take extra enzymes to finish the job.' Do more chopping on the rest.

          a a a a   a   a a a a   a a
          a a   a   a a a   a a

Of course, with two boys, my wonderful biochemistry lesson quickly deteriorated into all-out warfare. Jordan had some of the animals/ amino acids regroup on one side of the stomach and stage a counter attack against the enzymes. He said their mission was to escape into the intestines without being broken up. Matthew said the enzymes in the stomach were the ground troops. He sent a scout up the throat to call for reinforcements, 'We need help! Send down the protease enzymes to destroy the proteins. Bring some amylases to break the carbohydrates. Bring plenty of ammo.'

This went on for about 30 minutes with the enzymes being some bomber planes blowing holes in the enemy lines (the amino acid chains). I guess if you think about it, that is a pretty close description.

Where do enzymes come from and where do they go?

The three main groups of enzymes are:

- digestive enzymes
- food enzymes that occur in raw food
- metabolic enzymes

Enzymes are a part of every metabolic process in the body – from the working of our glands to the proper functioning of our immune system. Most enzymes have metabolic functions, however we will not be focusing on these as they have different jobs than digesting food.

We get enzymes to digest our food in two ways. Either we use the enzymes that our digestive tract secretes for this purpose, or we use the enzymes that occur naturally in all raw food. When you eat raw food, your body can use the enzymes in the food to digest the food instead of producing the needed enzymes for digestion itself. This is why digestive enzymes are classified as a food in the United States of America.

High heat destroys enzymes. Temperatures involved in cooking, as well as food processing methods, can inactivate most food enzymes. Each particular enzyme will have a different temperature stability range. However, most all plant and microbial-based enzymes are stable at 120 degrees Fahrenheit which is about the maximum temperature the mouth can withstand. So, if you can put the food in your mouth without burning, then the enzymes should be fine. Microwaving food creates a lot of heat. This heat can have an adverse effect on enzymes if the heat generated is high enough for a sufficient length of time. However, science does not know if the microwaves themselves directly have an adverse effect on particular enzymes.

Some fresh foods, such as fruits and vegetables, undergo irradiation at times in order to kill any bacteria, fungi, or other pathogens present. This process eliminates undesirable pathogens as well as increasing shelf life. We do not know whether this process destroys enzymes or not. Remember that enzymes are not biologically alive. They are molecules that facilitate biochemical reactions. Certain processes may inactivate enzymes by disrupting their molecular structure or damaging their active sites, but they are not 'killed' in the sense that most people refer to the killing of biological organisms.

Your body produces digestive enzymes throughout your lifetime, just as it does other cells, tissue, hormones, and other components of your body. The body also produces metabolic enzymes that run the biological processes that make life possible. Although some theories suggest you get a lifetime supply, one fixed allotment, of enzymes upon birth and you run off that initial supply throughout your lifetime, there is no proof of this, besides the fact it fights common sense. You might speculate that since enzymes facilitate all the functions in the body that they are a 'life force.' And you may further speculate that this life force diminishes as you get older. One might interpret this as a person's inherent genetic ability they are born with to produce enzymes at a certain rate, which may affect metabolism, along with other things. Metabolism and enzyme production does slow with age just as all biological processes slow with age.

When we eat a diet full of cooked and processed foods, we get very few of the natural enzymes from those foods. This means our bodies must produce the enzymes needed to digest the cooked food. Our bodies secrete digestive enzymes mainly from the pancreas and a few other locations in the digestive tract. The more digestive enzymes our bodies need to produce, the greater the stress is on the pancreas. Because the body needs to create these digestive enzymes, it uses precursors and raw materials that could be used for other functions.

Metabolic energy is also required. If you supply enzymes with the food eaten, either as supplements or in raw food, the oral enzymes will begin the break down of food while it is still in the stomach, allowing the pancreatic enzymes to do the 'finishing' work on food digestion. By consuming more enzymes, the body's resources can be put into growth, maintenance, and repair of the body instead of digesting food – functions people with physiological illnesses desperately need their bodies to perform.

Lack of enzymes resulting from eating a diet high in processed and cooked food has been correlated to many degenerative illnesses. Adding more enzymes to your system can help prevent certain disease conditions, help with healing, and keep the body running in tip-top shape. Of course, these factors can help extend your life and make it much healthier and happier. However, no amount of enzymes can give you eternal youth and ward off death forever.

We have two ways to increase the amount of enzymes coming into our body: by eating raw foods and by taking concentrated enzyme supplements. You would need to consume a really large amount of raw food to equal the quantity of enzymes found in a good enzyme supplement, and it would take longer to work. Taking supplements is more practical. Enzymes are a 100 percent 'natural' method of maintaining good health.

Enzymes in supplements come from either plant, microbial, or animal sources. Enzymes derived from microbial and plant sources have many advantages over enzymes derived from animal sources (pancreatic enzymes). Microbial and plant enzymes are active over a broad range of pH, temperature, and substrate affinities. Supplemental enzymes from microbial and plant sources are active in the pH range of 2 to 12. This means they can break down much larger amounts of proteins, carbohydrates, and fats over a longer time in both the stomach and small intestine.

Remembering our blender model, food may sit in the more alkaline upper part of the stomach for up to an hour before it enters the lower more acidic part of the stomach. Then it proceeds to the small intestine which is alkaline again (because bicarbonate is sent in to neutralize the acid).

Let's consider how different enzymes are affected by pH in the different phases of digestion. The digestive enzymes in saliva continue their activity in the alkaline upper stomach during the first 30 to 60 minutes after eating. The more food is chewed, the more saliva is mixed in, and the more these enzymes will be present. Swallowing food in chunks minimizes this interaction. However, the activity of these enzymes is limited to a pH level above about 5.0. The stomach acid destroys these enzymes when it arrives.

Enzymes from animal origins are destroyed by the low pH of the stomach acid unless they are specially (enterically) coated. If animal-derived enzymes such as oxbile and pancreatin are not enterically coated, they can work in the alkaline upper stomach, but are then destroyed by the acidic lower stomach. If they are coated, these enzymes will make it to the alkaline small intestine, but they will have no activity in the stomach. This coating may not dissolve in the small intestine in time to contribute much to digestion, or at all. The animal enzymes pancreatin, trypsin, and chymotrypsin are only active in the alkaline environment of the duodenum (Schneider 1985; Schwimmer 1981).

Pepsin is the enzyme activated by the hydrochloric acid in the lower stomach to break down proteins. It is only active in the highly acidic environment of the lower stomach once the hydrochloric acid is released. This may be up to an hour after eating. Pepsin is not active in the small intestine – too alkaline.

Enzymes derived from microbial and plant sources work at the pH found in the upper stomach where food may sit for over an hour. This is where a large percentage of the digestive activity of supplemental enzymes may occur. These enzymes continue to be active in the acidic lower stomach as well as later when the food passes into the alkaline small intestine. Although a few types of microbial and plant enzymes may be inactivated in the acidic lower stomach, they are not necessarily destroyed. They reactivate once again in the small intestine and continue working (Beazell 1941; Graham 1977; Griffin 1989; Lennard-Jones 1983).

Microbial and plant enzymes have a much wider variety of specific enzyme types as well enabling them to work on many more kinds of food. Pancreatin is a predefined blend usually from pig or cattle sources. While pancreatin is a blend of only protease, lipase, and amylase activities, microbial and plant enzymes offer protease, peptidase, lipase, amylase, glucomylase, alpha-galactosidase, cellulase, hemicellulase, invertase, malt diastase, lactase, pectinase, and phytase activities. Microbial and plant enzymes allow more flexible customized formulations and activity strengths.

Because the microbial- and plant-derived enzymes have these advantages over animal enzymes, you often need to take much fewer of these enzymes than pancreatic ones. The principal international standard for measurement is Food Chemical Codex (FCC) units. Enzyme activity should be assayed according to the FCC method, and the FCC measuring system should be used for measuring enzyme activity. Enzyme activity is the important measurement and not weight units such as milligrams. Since there is no direct relationship between weight and activity, weight measurements do not convey any information on how active the enzymes are. Low-activity enzyme products may weigh as much as high-activity ones, and fillers may add to the weight of the product but not to the activity.

If I have 100 milligrams (mg) of active enzymes and I cook it in the oven for four hours completely destroying any trace of activity, I still have 100 mg of inactive enzymes in the substance. It just will not do anything – like a dead carcass. I can also have a capsule with 10 mg of active enzymes and 90 mg of filler which equals a total weight of 100 mg, but this would be a far different product than having 90 mg of active enzymes and 10 mg of filler. Always look for, or ask for, activity units, preferably the authorized FCC units.

After digestive enzymes complete their task of breaking down foods, a few things can happen. The stomach acid might denature or destroy the enzymes. If this happens the enzymes will be broken down into their component amino acids and these used as raw building materials in the body. Proteases, amylases, and other enzymes may be absorbed into the bloodstream where they may go about the business of cleaning out foreign particles. This includes removing free radicals, potential allergens, reducing inflammation, and transporting waste out of the body (these are discussed in other sections of this book).

We now know that some digestive enzymes are recycled in the body, in much the same way as the bile salts. The body reabsorbs the bile salts and enzymes at the end of the small intestine. They make their way back to the liver and pancreas, and are used again. This indicates that supplementing enzymes regularly can have a cumulative effect on good health (White *et al* 1988). This may be one of the contributing reasons why health continues to improve even after enzyme consumption stops. It also makes enzyme supplements an extremely good investment because they may continue to work much longer than just one trip through the intestines.

At other times, the body treats enzymes as the proteins they are. Enzymes do experience some wear and tear as they work. A little strain here, a little bump there. The active site is nudged and pulled and deteriorates bit by bit until it is not a true functional template. Enzymes 'age' over time as they work, and after a while they are not in the prime condition they once were. Eventually, they are simply broken down into individual amino acids by other enzymes. Enzymes have an affinity for denatured protein (sick or dead tissue) so they selectively break down deteriorating enzymes while the 'healthy' ones go about their work. The body then uses those amino acids as building blocks just as they would use amino acids from other sources. Different enzymes have different 'lifespans' depending on type and how often they are used doing work in the body. Some enzymes may simply be excreted from the body.

Some people wonder if their body or pancreas will stop producing its own enzymes if they start taking digestive enzymes regularly. There is no evidence consuming enzymes even over a long period of time adversely affects pancreatic function or response, nor that internal enzymes were not produced once supplements were stopped (Friess *et al* 1998). Digestive enzyme supplements may even improve your natural enzyme production (besides the previously mentioned routes or giving the pancreas time to recoup). Certain enzymes are produced by the intestinal lining. With any type of gut injury these can become unavailable. Taking enzymes can facilitate gut healing, thus assisting the mucosa and villi cells to re-grow, which allows the intestinal enzymes to become functional once again. Then the oral enzymes may not be as necessary; perhaps not needed at all. This is consistent with many finding they need less enzymes over time, not more.

Enzyme safety

Enzymes are some of the safest supplements available because of a fundamental characteristic they all have. They have very specific mechanisms that are well known, and they do nothing other than their particular job. Enzymes work according to ordinary biochemical properties and are not living organisms complete with free will and instincts. A protease will not be halfway through breaking protein bonds and decide to go over and start breaking down sugars or carbohydrates. A digestive enzyme with the function of breaking down fats will not decide to hop in the bloodstream and wander up to the lungs to see if the metabolic enzymes need any help with respiration. Enzymes are one of the most well-studied and characterized compounds in biological systems. Because of this, enzymes have far fewer side-effects and unknown possible reactions than other compounds, supplements, or medications. This makes them extremely safe to take for very long periods, even for a lifetime.

Consuming too many enzymes is practically impossible to do. No known toxicity has been demonstrated at any level of enzyme dosing in animal studies or in humans. Animals survived outrageously large quantities of enzymes without damage so it has been impossible to find a lethal dose. Researchers fed guinea pigs and rats a daily enzyme dose for six months which would have corresponded to around 250 tablets per day for a 60 kilogram (134 pound) person. No ill effects. Rats were fed enzymes equivalent to a human dose of 2500 tablets daily for a short period and the rats only seemed a little fatigued (Lopez 1994). Research has also looked at cell changes and mutations. They found no negative affects at all.

There is a wealth of recorded information on the safety of using digestive enzymes for the past 70 years, going back to the early 1900s. This would be both reasonable and understandable to expect because, after all, our digestive system is swimming with digestive enzymes all the time anyway. If we don't have enough enzymes regularly, that is when we get sick. Every time you eat fresh fruits, vegetables, or other raw food you are consuming digestive enzymes. Eating fresh foods is even something highly recommended for good health throughout our lifetimes. There is far more evidence showing you may very well become ill if you do not take enzymes than if you do.

Digestive enzymes have been used safely for decades and are classified as a safe food in the United States. Enzymes have GRAS status (Generally Regarded As Safe). Since they are not in the realm of the Food and Drug Administration (FDA), specific digestive enzyme blends are neither FDA approved or unapproved, just as oatmeal is not specifically FDA approved or unapproved for you to eat.

However, enzyme therapy is currently FDA approved in the treatment of certain health conditions (Lopez 1994). This is similar to the FDA declaring eating oatmeal may reduce the risk of heart disease. FDA-approved enzyme therapies include treatment for:

- cardiovascular disorders
- gastrointestinal conditions, particularly pancreatic insufficiency and related disorders
- replacement therapy for specific genetic disorders and deficiencies
- cancer treatment
- debridement of wounds (degradation or cleaning out of dying or dead/necrotic tissue)
- removal of toxin substances from the blood

As you can see, many of the specific biological issues identified with neurological conditions, including the autism spectrum, fall into one or more of the approved categories for enzyme therapy.

Several governmental regulations and agencies have the job of regulating enzyme production for quality in the United States. These include the United States Department of Agriculture, the Food and Drug Administration, and the Environmental Protection Agency. Enzymes manufactured for use in food must follow the FDA's current Good Manufacturing Practices (cGMP) and meet compositional and purity requirements. The Food Chemicals Codex (FCC) is a compendium of food ingredient specifications developed in collaboration with the FDA. The FCC has established certain units of activity for enzymes to ensure the quality and potency of a product. Look for these official FCC units or their equivalents on the label.

CHAPTER 7

Unlocking the Treasure Chest

After being on a casein-free, gluten-free diet for two months, I realized we had a problem with dairy. The tiniest amount of casein would create major emotional meltdowns, migraines, inability to transition, and whining.

However, my family was having quite a dreadful time adjusting to this diet. I missed my pasttime of baking traditional breads, cookies, and muffins. Restrictive diets are no fun unless you are deliberately trying to avoid lima beans. We found the diet was practically impossible to do as well. It was always one thing after another you could not have, or a place you could not go, or the product was contaminated, or the formulation changed. This diet required 100 percent compliance. There was no room for anything less than perfection, absolutely no 'cheating' or 'infractions' at all; 98 or 99 percent was not good enough. You were commonly advised to stick with it for one year, with an estimated one in three people maybe seeing improvement.

My boys were seven and nine. One of the worst aspects of restricting so many common foods is that my boys were in elementary school and wanted to eat school lunches as well as do other things with their friends. It was just one of many such problems that sprang up. Then, there is art class and other projects involving glues, markers, creative clay, and other potentially problematic stuff. These substances may contain gluten that could be absorbed through the skin or inhaled.

For snacks and party invitations at friends' houses, I needed to make special arrangements with the host and send different foods, if I could even find out in time. There were no more spontaneous stops to enjoy a pretzel during shopping trips. Even though the diet helped us in some ways, it effectively created an entirely different set of problems. Food is a very big part of many social interactions in most cultures. Socializing is often a challenge for children with behavioral problems, sensory issues, or autism conditions as it is, and having to make an issue out of food everywhere just made it worse. Avoiding almost every common food and substance was making life harder.

Then there is the expense of it all. Five dollars for a cake mix?! I am used to paying 69 cents. Four dollars for a bag of chocolate chips when I used to pay one?! Everything was three to five times more expensive. I could order special foods on the Internet, but the shipping costs were often as much as the food. I could no longer take advantage of local weekly sales for common foods and household supplies. We had to take special food everywhere we went. I do cook and tried many, many recipes. Not much of it tasted very good to us.

The nutritional supplements were at least $15 a bottle, often just for one child, and I heard many people saying they gave their child a couple dozen supplements each day. That comes to around a good $300 a month for one child, if I take the conservative route. From the majority of the conversations with other parents, I definitely wasn't the only one struggling to make this workable or affordable.

I can understand if you have toddlers and preschoolers it might be more manageable because you have more absolute control over them and their environment. But when children get older, it can be very different. A nine-year-old does not want to bring his own 'special' cupcake to a birthday party. I kept thinking, 'There has *got* to be a better way.'

In one of the Internet groups, I heard about digestive enzymes and how you might use them for the breakdown of the foods people were trying to eliminate. Cindy Kelley was among the parents constantly asking questions on the casein and gluten peptide issue and digestive enzymes. At the time, there were many people talking about a new product about to be released called Peptizyde. This product supposedly was designed to break down the theorized potentially offending gluten, casein, and other proteins. There was speculation that Peptizyde might let you eat gluten and casein foods occasionally, or even regularly, without having a negative reaction.

It seemed to be a much more efficient way to accomplish the same purpose of eliminating the suspected peptides, without having to continually remove foods and then add in supplements. Plus, we would be getting nutrients from our food instead of supplements, and it could facilitate good digestive health along the way. I realize a casein-free, gluten-free diet has helped certain people, but I really did not want to start down that slippery slope of having to continuously remove one food after another. Nor did I want to be one of those whose health got worse on the diet. I certainly respect anyone's decision or need to do a restrictive diet. However, I felt using the enzymes would be the healthiest alternative I could provide my family – that is, *if* the Peptizyde actually worked for us.

Part of doing anything is your own personal comfort level and lifestyle choices. I have a background in agricultural science, and enzymes have been a standard part of agriculture for some time. So my comfort level with enzymes was way-high to begin with. Because I also knew how nutrition and biology worked in plants and animals, my comfort level with eliminating entire food groups was way-low. Each person needs to evaluate what is best for their situation.

Supplementing with enzymes can be very beneficial when added to a restrictive diet because they can pick up hidden ingredients not disclosed on the label and other sources of contamination. Enzymes may allow planned infractions for special occasions. I thought a restrictive diet would be soooooo much more manageable if the boys could eat a school lunch every now and then with their friends. Or go to birthday parties and other special events. Even if enzymes only allowed you to have trace amounts of casein or gluten, that would significantly open up the range of food possibilities, and unload a horrendous stress from the family.

Like most people, I wondered how long you generally had to stay on a casein-free, gluten-free diet, should it prove helpful. Most people believed it was for life because of the lack of information on this theory over the long-term. Later, I heard a figure from an organization specializing in information on the casein-free, gluten-free diet. They estimated that of all the people who saw success on this diet and then went off it after some time (with 'some time' not being defined), one in five can successfully leave it without problem, but many others regress. So after some time, around 20 percent would be fine if they reintroduced casein and gluten. Maybe a leaky gut healed? But there was no way to know who was going to do well and who would go crashing back into problems. Often a person would leave the diet and everything would be going fine for a while. Then maybe after the second month or so, Wham! – a major relapse. There wasn't a way to tell who might be affected in this way. No one is even really sure what causes the relapses.

While waiting for Peptizyde to arrive, I bought a bottle of protein digestive enzymes from the health food store to see if that would work in the meantime. Disaster. It did not! We could tell within hours of each dairy plus enzyme trial. About two to three hours after ingesting any bit of casein, I would have a migraine, Matthew would be whining and banging his head, and Jordan would be flopping around the house complaining of a headache. So I know what it looks like when the enzymes do not work on our specific problem. All enzymes are not equal and we apparently needed to try those specific ones to see if they would do the job.

The making of Peptizyde – The story behind it

As soon as research discovered certain partially-broken-down proteins might be causing problems, the logical thing to do was to look for a digestive enzyme to break them down. There was a long search and several attempts at developing such a product. Finally, National Enzyme Company was contacted, where Dr Devin Houston was head of research and product development. He accepted the challenge.

As it turns out, Dr Houston had a unique background that made him a particularly good scientist to work on this problem. He had done extensive research with enzymes, hormone receptors in the brain, the human response to opiate triggers, and worked with autoimmune immune and neurological conditions at a few prestigious medical schools, with numerous scientific research publications. Dr Houston applied the same rigorous research methodologies to the development of enzyme products. How interesting that this all seemed to be just the right fit for this project! His experience and credentials were all very suited for the challenge. But the question remained: Could he make an enzyme product that would be practical and effective in everyday life, possibly eliminating the need for a restrictive diet?

Dr Houston was able to formulate a product. Known as SerenAid, it became available in 1999 from Klaire Labs. Unfortunately, some changes occurring during this time led to some disruption among the many individuals involved in this collaboration. In March 2000, Dr Houston left NEC to work as a consultant and researcher.

A short trial was conducted with SerenAid early in its development The parent-judged results showed that in four weeks on this enzyme product participants improved somewhat in 13 rough categories (eye contact, socialization, attention, mood, hyperactivity, comprehension, speech/language, sound sensitivity, digestion, sleep, ritualistic behavior, anxiety/compulsions, stimming). However, many families found the product a bit expensive for regular use and it did not allow the reintroduction of casein and gluten for most people.

During this time, other companies looked into possible enzyme products designed for people with autism conditions. Another investigation similar to the one with SerenAid was tried involving a product called EnzymAid, only this time over a 12-week period. It was also parent-judged, not blind or with placebo, but again showed some improvements in the same 13 general parameters. This trial involved 22 children with a variety of autism spectrum and attention deficit conditions. Over half or the participants were not on any type of restrictive diet.

The manufacturers, researchers in the autism community, and parents who tried all of these enzyme products as well as other enzyme products available adamantly recommended these be used with a 100 percent strict casein-free, gluten-free diet. They were not for even an occasional break from the diet. Many felt a more effective enzyme product was not even possible. The interest in developing enzyme products waned, research dropped off, and attentions turned to other things and different avenues.

Devin Houston still believed it was possible to develop a better enzyme product for the suspected peptide problem. Although others had given up, for him, the hunt was still on. Maybe he was struck by what the struggling families had to go through. Maybe he sensed the improvement that was possible. Maybe it was the challenge. Working on his own, he was determined to create something better. In April 2001, he released two new products under his own label, Houston Nutraceuticals Incorporated (HNI). One was Peptizyde for the breakdown of gluten, casein, soy, and other proteins. The other was HN-Zyme Prime (Zyme Prime). Zyme Prime is a robust all-purpose, broad-spectrum product that contained a range of enzymes designed to break down all food groups.

What makes Peptizyde so special? A particular enzyme is identified with a unique function in the breakdown of casein proteins. The enzyme is known as dipeptyl dipeptidase IV, or DPP IV. DPP IV occurs naturally in the intestinal cell lining, but this enzyme along with others in the gut may decrease if there is inflammation or injury to the mucosal lining. You can also find DPP IV in the secretions of proteases from certain fungi, the same used to provide oral enzyme supplements. Peptizyde contained the highest amount of this specific enzyme of any product, along with other proteases needed to help it work more efficiently.

Research shows a synergistic effect occurs when DPP IV combines with two other kinds of proteases (Byun *et al* 2001). Peptizyde contains all of these enzymes in one capsule. Gluten and other bonds in casein may be broken down with products containing strong general proteases (Doumas *et al* 1998). Developing enzyme formulations is as much an art as a science. Other work with enzymes verify this as well – certain mixtures are very effective, whereas giving the individual enzymes separately did not produce the same results. This is why one formulation may work quite well for one person but not so well for another, depending on individual conditions, sensitivities, and biology.

What caused the ruckus – Improvements on Peptizyde

In a matter of days of starting the Peptizyde and Zyme Prime enzymes, some of the parents noticed immense improvement in their children's behavior and physical symptoms. In addition, a few people decided to try these enzymes with casein or gluten. I was one of those parents.

The day I received the Peptizyde, I opened the bottle and gave my boys one capsule each and an ice cream sandwich (I felt that even if the product did not work, at least they could enjoy the experiment). I joined in on this trial, too. Nothing happened. Three hours later, nothing happened. By three days later, nothing happened. No migraines. No whining. No nothing. I was never so glad to see a product produce absolutely no visible results!

So again and again we repeated this, testing Peptizyde very vigorously. I gave enzymes with dairy and grain food at every single meal and snack around the clock for a week. I wanted to know what the upper limit was, but I never found it. Our dietary 'infractions' were essentially continuous with no return to any of the negative symptoms we previously saw with casein or using other enzyme products. And so, we were soon off the restrictive diet by default. It was wonderful!

Actually, we saw a great deal more improvement than when eliminating foods. I switched to these enzymes right away and went back to a well-balanced diet including all foods regularly. Our food bill dropped. The boys could participate socially again. The aroma and tastes of traditional baked goods filled the house. We were very happy campers.

One by one similar messages came from other families as the parents tried these enzymes. Kids were getting better. After a few nervous weeks, it became apparent that these enzymes were working. They were really working! The messages came littered with:

'My son is starting to talk. All I started were the enzymes.
 Is this possible? Am I imagining this?'

'My daughter is sleeping through the night for the first time.'

'My son has stopped his compulsive behaviors and stimming.'

'My son's therapist said my little boy had his best week ever today.'

In addition, it was apparent that people who started using Peptizyde while remaining consistently on a restrictive diet were also seeing significant improvements. Improvements came even if dairy and gluten had been rigorously eliminated for over one or two years.

One typical tale is that of a little four-and-a-half-year-old boy who had been on this diet for around two years until the family started these enzymes. They were thinking of leaving the restrictive diet anyway. When they started the diet, the mom confessed having such great hopes for recovery. But after two years of depriving him and watching him like a hawk, writing everything down in notebooks, everything he ate or did not eat, trying to find patterns or figure out what kept him up all night this time, or why he was being too hyper, they had had enough. She described how she and her husband had just accepted the autism and found a way to focus on the blessings (and there are many).

After a month of using these enzymes and a non-restricted diet, she noticed a significant increase in her son's talking and communication. Her son also showed much more awareness of his surroundings. At first, they were giving the enzymes just at meals but then started giving the enzymes in a drink throughout the day (an idea used by some), and they saw a huge improvement. She confessed she actually felt like entertaining and trying new recipes again. The enzymes lifted a huge burden from their life. These enzymes were proving they were not just a one-for-one swap or substitution for food elimination. These enzymes were doing much, much more. This appeared to be a major turning point.

Another parent wrote me that after using these enzymes for over four weeks, they saw an increase in their son's multi-level communication, singing songs he made up himself, and potty training. They noticed he understood more complicated concepts like cause and effect, he significantly reduced his tantrums, he talked about what he did in school (never before), he significantly reduced his watching of videos, and correspondingly increased his imaginative play with toys and play with his little sister. Other new developments including helping his mom with cooking, the laundry, and even cleaning! He also started imitating prayer at dinnertime, and singing a group song at church with his class.

A further interesting find was this family was able to discontinue giving many of their previous supplements. After a bit, they could not understand continued belligerence, tantrums, and hyperactivity. They wanted to rule out once and for all if it was the enzymes. It wasn't. Once they removed all additional supplements most of that negative behavior went away. The mom described seeing an amazing child who still has issues but actually says hello to people, enjoys playing with other children, loves to do crafts in school, and who is going to be ready for a school life that will help him to keep working on his primary issues of social skill struggles, large groups and noises.

Not only were some of us able to leave a restrictive diet if we wanted to, we had opened the treasure trove of benefits from enzymes. The rewards of gut healing, immune system repair and support, fighting pathogens, and correcting nutrient deficiencies were spilling out into our lives. The improvements went far beyond being able to add a few foods back. We were seeing gains and new skills in our children and in ourselves that were far more significant. We didn't know why at the time, but as one joyful message after another sprang up, it was evident that something remarkable was happening.

However, even though many changes in life are quite positive, change is still change. And change can be hard. Not everyone embraced these new improvements. There was instant alarm among some upon hearing any discussion of intentionally deviating from the strict diet for any reason. After all, other enzyme products had not permitted leaving restrictive diets. Some people were adamantly and actively against this new ability to deliberately consume casein and gluten, even occasionally. The traditional position on this diet required 100 percent compliance, no exceptions any time for any reason. Cindy, I, and some other parents met with hostile replies from some groups when we mentioned our children's successful improvements. People upset that kids were getting better? That our children were improving so much with a relatively simple and inexpensive supplement? Wasn't that the point of all this?

There was some skepticism on whether or not these improvements would last. To a certain extent, this was a legitimate concern. Often enough, someone would be doing okay on a casein-free, gluten-free diet for some time, and then test the waters by going off the diet. Sometimes things would go well initially, but after a month to three months of consuming casein and gluten regularly, a relapse would occur, often in a major way. Because of this history, many people were tempering their enthusiasm until time proved the improvements with these enzymes would last. And so the waiting game began.

Many parents were seeing fantastic results and felt it was their perogative as well as responsibility to try to better their child's health and life by whatever means was available. Many adults wanted to keep the same benefits for themselves.

Dr Houston stated his position on this matter in the beginning and it continues to be that the purpose of digestive enzymes is not to detract from the usefulness of any particular diet in helping people. High-quality enzyme products such as Peptizyde are meant to achieve the same purpose as that of the casein-free, gluten-free diet: to reduce the amount of any potentially harmful exorphin peptides produced. Based on well-characterized mechanisms of enzyme actions, one may assume that supplementation of a diet with Peptizyde helps to not only reduce or inhibit the production of potential peptides from food proteins, but supports digestion, insures complete degradation of food proteins such that they are not allergenic, and increases bioavailability of food proteins. Due to the established safety and non-toxicity of enzyme-containing supplements, we believe there is sufficient historical, anecdotal, and scientific evidence present from the use of hundreds of enzyme products to justify such experimentation in a prudent and responsible manner, and should be allowed to those who desire to try this alternative.

This was a position I could accept and work with as a parent and as an individual suffering from a pervasive neurological condition. It was practical, cautious, and yet open to looking for progress and hope. With the success of these enzyme products and enzyme therapy with autism spectrum and other conditions, many people are now adopting a similar point of view.

However, this was not a position some others were prepared to accept. Some individuals complained that enzymes that allowed you to regularly consume casein and gluten were not possible so it couldn't be working, or that the product needed rigorous double-blind controlled testing before being used.

There was a bit of a problem with this line of proving the effectiveness scientifically. It seemed very selective. I have seen very few clinical tests on anything in the autism biological treatment arena, especially regarding nutritional supplements or therapies. Where were the rigorous trials for the other measures promoted specifically to help the physiological concerns in autism and related conditions? Even restrictive diets can be risky depending on the individual. Studies seemed few and far between, if done at all. Routinely, suggestions and ideas floated among the searching parents without anything at all except maybe a friend passed along an idea that helped someone else, so you went with that.

A mountain of studies had already proven enzymes were extremely safe in general, along with FDA approval for various conditions. Often, physicians recommended enzymes to their patients with autism or neurological conditions to improve eating patterns, bowel regularity, and intestinal health. So this was not entirely new. Other researchers and business affiliates had become uninterested in the enzyme angle choosing to leave it behind. Maybe because they were convinced it could not be done, or perhaps there were far more lucrative alternatives.

Studies do not necessarily offer definitive solutions even when conducted. Sometimes you get studies arriving at completely opposite results. And even if you had many studies to draw on, none of them could guarantee that your particular child would benefit or not. There is always a balance between science, safety, and practical solutions.

Another glaring issue was that kids were continuing to improve. They improved by leaps and bounds. They improved no matter what kind of diet they followed. They improved in a wide number of areas and symptoms. The reports on improvements were popping up like dandelions after a spring rain.

And, the children usually improved very quickly. After tracking results for a while, we found that about 50 percent of the parents reported they saw noticeable improvement by the end of the first week. By the end of the second week, that rose to an impressive 80 percent. We now know that most everyone can tell if taking these enzymes will help or not by the third to fourth week. That is a pretty efficient avenue. And it is fairly inexpensive compared to other alternatives or therapies available at the time. You can usually try enzymes for about three weeks and under $50.

The parents seeing their children improve were not about to give this up. Some children were talking for the first time ever. Or sleeping soundly through the night. Or had stopped having tantrums throughout the day. They were playing with brothers and sisters and doing unusually well in therapy sessions. Adults like myself were seeing migraines and other pains disappearing, their digestion improved, and their overall health was better.

Some told the parents they did not know what they were doing, they were not intelligent enough to make such decisions, they were confused and should give up this crazy idea. I was one of those parents. And I did not much care for the implication that I was not qualified to observe and evaluate the remarkable effects these enzymes were having on my children or myself. Besides having lived this unpleasant journey with both of my children, I had lived this nightmare myself. I knew firsthand what is was like.

A great many of the parents were very well educated, thoughtful, and resourceful. They had been working with their child's condition for a long time. The background, the science, the pros and the cons had been discussed, researched and re-researched until the decision was made to try this alternative in their family. Several other parents had tried earlier enzyme products. We knew the risks and potential benefits. All the solutions for neurological conditions were offered this way. Every supplement, therapy, and medication was simply a series of 'Try this and see if it helps.' Then, 'No, didn't work? Well, then, try this other thing.' The feeling was that enzymes were no different from the other therapies, medicines, supplements, and diets we were encouraged to try.

It's not like there was a set of useful guidelines to follow or significant other alternatives. Often, parents are given no hope or just a series of hit-or-miss attempts at trying to find something, anything that would help their particular child. When dealing with such a wide spectrum of conditions, each person has to find whatever might help their individual situation. It's just the reality of dealing with a elusive, ill-defined condition. Insurance usually covered very little, if any at all, of the doctor's fees, multitude of tests, the ton of supplements, or special foods. Parents were looking for something new because other things were not working, or not working well enough. And now parents had these new enzymes, this Peptizyde stuff, this inexpensive, safe, natural item that was, in a number of cases, bringing their child to life.

The tension was intense among the parents and members of the various groups. Tempers flared. Each 'side' felt they were supporting the best and safest course of action for their own situation. Maybe so. There is no one definitive path for everyone.

Suspicions were thrown out. Maybe we were being tricked. Maybe we weren't telling the truth, or were paid imposters. The parents were told they needed to listen to 'the experts,' but there was no consensus on who that was. Many of the 'specialists' would fuss amongst themselves over who really knew what they were doing and which theory was best.

The parents using the enzymes countered with speculation that those with a vested interest in insisting people stay on the restrictive diet seemed to have a lot at stake. There was, after all, a great deal of money spent on diet foods and supplements. If enzymes made all that unnecessary, someone would be out a substantial amount of revenue. Anyway, why were people upset that kids were getting better? Wasn't the idea simply to get kids as well as adults into better health as effectively as possible?

In an attempt to defuse the situation, someone quietly did a very thoughtful thing. The Enzymes and Autism group was established on the Internet as a seperate forum for people to learn about and discuss enzyme therapy. A place designed for education and support of digestive enzymes in practical everyday use. Those interested in enzymes now had a 'spot' of their own, a harbor away from the tension, and a place to explore their new treasures. A small group of parents drew virtual straws, and I lost. That is how I became listowner and moderator. Feeling sorry for me, Cindy agreed to be co-moderator. That's Cindy, always there to help someone in need!

The messages of those using enzymes poured forth with hopefulness and happiness, as improvements littered the messages. It was like Christmas with everyone opening their gifts exclaiming what gems they were enjoying at their house.

'My daughter is using the potty!'

'My son asked to help with the dishes!'

'I could make my daughter the birthday cake she has wanted. It was a wonderful day.'

'The teacher said my son sat attentively all day. The red rash and dark eye circles are gone!'

'We enjoyed a regular family meal at a restaurant! No stress, no reactions. We are so happy.'

My son's experience was just one of many. Matthew 'blossomed.' He has always been the more severely debilitated of my two boys. His eye contact is great. He is no longer continuously irritable, angry, argumentative, or highly anxious; just consistently very agreeable and pleasant. He readily does his homework and helps around the house. Socializing and interacting with others has taken the place of most of the rocking, bouncing, and other repetitive behavior. Matthew is very attentive and aware of where he is, who is around, and what is going on around him. He used to wear out quickly, but now has lots of energy. It is as if he is now soaking up life with every breath. Before being casein-free, Matthew would have screaming fits when he had to change activities and he wasn't ready. Eliminating dairy got him to transitioning with clenched teeth and a few tears instead. Now with these enzymes, Matthew transitions with ease. No problem. He will just jump up in a happy mood and go. And the head-banging has virtually stopped. Matthew is finally functional in life. A primary goal for nine years finally achieved.

Jordan is also doing quite well. Little Bear rests on the bed during the day. Jordan only twirls Little Bear's green ribbon going to sleep at night. His eating is significantly better and he has much more energy (Jordan, that is, not Little Bear). Both the boys' teachers commented to me separately about seeing improved awareness, attentiveness, getting assignments done promptly, and just overall pleasantness after three weeks on enzymes. Their Taekwondo instructor remarked on both of them being 'more attentive' and 'putting in better effort' and 'really progressing now.' Life is good.

The waiting game

And so, we waited to see if the improvements with Peptizyde would hold fast for at least three months, the time when most people experienced a 'crash,' (an abrupt negative regresssion). Getting to four months was even more definitive because most everyone that regressed did so certainly by the fourth month. Caution is good, and everyone has their own comfort level with different things.

Cindy's son was eight years old at the time and previously met the criteria for Asperger's Syndrome. He had been on a restriction diet for over one year. It started innocently enough as the casein-free, gluten-free diet. However, like many people, she found herself in the downward spiral of having to continuously remove food after food after food to maintain her son's level of health and disposition. After starting the new enzymes, she was able to restore almost all foods to her son's diet. She talked about his immense improvement with the enzymes that she did not see with the diet alone. She was having one successful 'infraction' after another, until one day, she realized she was not restricting any kind of food at all.

Because Cindy's son had been on a restrictive diet for so long, I knew more people had their eyes on her, just waiting to see if her son would crash. At about the second month mark, I asked her if she was worried about her son relapsing. Her answer was very insightful:

'If my son has a major collapse, so will I! Actually, I won't. I'll just rack this up to one long, beautiful vacation. I don't really foresee a collapse, and really don't see this as the same thing as 'going off the diet' as others have done with possible regression. The reason is my son is better than ever, and improved even more with these enzymes. I think they are catching foods I was not even aware of before, even though he was down to very few foods. Other people maintain their current level of health for a while, and then gradually decline. My son has improved significantly since starting enzymes and going back to all foods.'

Cindy's point is that when people have reintroduced casein and gluten without enzymes before, or even with other enzyme blends, they would maintain the current level of improvement and then regress. Rarely were there reports of kids springing to life or showing tremendous gains when going off-diet. With these enzymes, she saw such immense improvement she felt significant healing of some sort was going on. As it turned out, she was right, in a major way.

Biological systems are complex and unique. Although a peptide issue (or any food issue) does not affect everyone, it is so little understood, no telling what would actually happen.

There was a rally among some of the parents to start keeping records on improvements, reactions, side-effects, etc. Having little guidelines or data to help you in testing alternatives was very common in the myriad of treatments and supplements available. We tracked any and all results members gave – positive, negative, or no change at all. The feeling was that even negative results were important because we might learn which people were more likely to benefit from enzymes, or discover better ways to give them. Knowing when not to use a therapy can be just as important as knowing when to use it. The idea was to develop as many guidelines for using enzyme therapy as we could; to contribute a bit to ending the giant chaotic free-for-all of alternatives.

Some parents continued to give enzymes with casein and gluten regularly, some allowed casein and gluten only occasionally, while others kept a strict diet. We embarked on discussions and learning about enzyme therapy; helped each other troubleshoot problems and supported one another on this new adventure into health and hope.

So, while we waited, I tracked results. I was pretty comfortable with this having done research in biological sciences. Many of the parents and other individuals volunteered their findings and talents as well. And a number of us researched why things might be happening. The fact that both my sons were doing considerably better was huge for me, to say nothing of the fact I felt so much better taking these enzymes myself.

So what great things was I seeing? I suppose I could characterize it best by saying what was not happening. With Peptizyde, my daily migraines are almost gone. They were about 50 percent gone with the medication amitryptiline, and now they are up to around 85 to 90 percent gone with enzymes. My husband says I am not so irritable, overly sensitive, and picky about everything (thanks, darling!). He says I am much more social, happy, and conversational. My kids say I don't fuss so much at them. I do not have to take a nap after a two-hour meeting with clients. I have so much more energy – it is like one day now is equivalent to three or four days before.

My sensory problems have dropped considerably (although not altogether). This translates into a huge improvement in daily life functioning. I don't have to use every ounce of energy to get out of bed in the morning. In fact, I can just get out of bed, go through an average day, and go to sleep without much difficulty at all. And sleep is restful. It is no longer a disturbed painful time. Awesome! Another of my dreams achieved.

Getting good sleep has such a big impact! I strongly believe fixing a sleep problem is a major priority, just as my neurologist expressed.

It took a little bit for me to adjust to feeling better. I suggest to other parents that this may be some of the adjustment that their kids may be feeling. They feel good, but feeling good is different, so they have to adjust to this different feeling. I would go up the steps in my house and stop. Hmmmmm. There was no ringing in my ears and pain in my head as before the enzymes. So, I would try going down the steps, then stop. Yep, same thing. No extra pain in my head from the motion. Cool! So, I would do it again and again. Almost practicing going up and down and learning what this feels like without that rattling pain. If I was a parent watching my daughter do this, I would probably rush over to the computer or phone saying, 'My daughter has started on enzymes and seems to have this new behavior of repetitively going up and down the steps. What should I do?'

Then there was the feeling-good-all-day reaction. Before the enzymes, I would just run out of energy having to deal with constant head pain. I would be wiped out by six in the evening. Consequently, I did not achieve very much late in the day. I had semi-consciously been carefully managing my energy so as to get everything done earlier in the day. After starting Peptizyde, I was completely functional and feeling good, even as late as ten o'clock at night. So I would sit in a chair and, being incredibly aware that my head did not hurt, think about what I should do next. What should I do now...now that I have four extra hours every day?! Four bonus functional hours every day to do whatever I needed or wanted to do. It was a little strange to say the least. Overall, it took a good six weeks for me to really adapt to feeling better. I spent many of those new-found hours tracking trends in the enzyme group and researching enzyme therapy.

Our discussions on enzymes included many aspects of gastrointestinal health, other supplements, diet and nutrition, and coping with neurological conditions in general. Truly, a lot of the credit for the newer findings in enzyme therapy for autism conditions goes to the group effort of the parents and other adults working together. It is a caring, thoughtful group of people pulling together and combining resources to find practical solutions to help not only their loved ones but others as well.

During this waiting time, some other organizations got busy in this area of enzymes ...really busy! A renewed interest in enzymes for autism spurred new analysis of existing lab reports and ideas. New products were in the works focusing on different aspects of gastrointestinal problems and health.

The wheels of change were turning.

CHAPTER 8

Enzymes and Disease

Although the use of enzymes with autism spectrum and neurological conditions is rather new, the use of enzymes with other conditions is not. In fact, the whole idea of using enzymes as a therapy for health is about as old as life itself...literally. Animals licking their wounds would be applying a steady supply of enzymes through their saliva. You can find references to enzyme therapy quite extensively throughout human history.

There is a great body of research on enzymes and various diseases. A sampling of the literature is cited in the reference sections at the end of this book, but there is much more. What is firmly established is that enzyme use has been going on for a very long time, very safely, and very successfully. Enzymes fight disease through a variety of mechanisms, which makes them an ideal therapy for many biological conditions (Stauder 1995).

One of the more 'modern' researchers into enzyme therapy was Dr Max Wolf, born in 1885 in Vienna, Austria. He was a Professor of Medicine at Fordham University in New York. Dr Wolf became aware of the key role that enzymes played in the vital processes of life. He was one of the first to envision the therapeutic possibilities with a better understanding of enzymatic actions. This went on to become his life's work. Dr Wolf's original enzyme formulation, Wobenzym, has been administered successfully for almost 40 years, with mountains of research to its credit.

Starting in the 1930s, Dr Edward Howell conducted numerous studies on the effects of enzymes in digestion and health. In one study, rats fed a diet of cooked and processed food lived about two years, while the rats that ate raw food lived about three years. The diet of cooked foods resulted in early death. He has also noted that the rats eating the cooked food showed a decrease in their brain weight while their body weight went up. The book *Food Enzymes for Health and Longevity* by Dr Howell lists over 400 scientific studies on the value of enzymes for improving health from 1904 through 1938. Reading through this literature review of these 400 studies gives a lot of insight into how the knowledge base of enzymes has evolved. Since there is such a substantial proven history for enzymes, much of the basic types of studies are no longer deemed necessary.

Dr Francis Pottinger wanted to determine what happens to the body when primarily denatured, incomplete, or processed foods are eaten continuously. The Pottinger Cats Study lasted for ten years, with three generations of cats being studied. Approximately 900 cats were involved in all. He took two sets of cats and fed them only raw milk and raw meat (high in enzymes). He took three more sets of cats and fed them cooked meat and pasteurized milk (no enzymes). The cats eating the raw food were disease free and healthy generation after generation (Pottinger 1983).

However, the cats eating the cooked and processed foods were not in good shape at all. Each succeeding generation was worse than the previous one. They developed a wide variety of 'modern' ailments, including heart disease, cancer, kidney and thyroid disease, tooth loss, arthritis, reduced bone mass, difficulty in labor, diminished sexual interest, infertility, and irritability so intense that they were dangerous to handle. By the end of the first generation, the cats started to develop the same degenerative modern diseases humans get and became quite lazy. By the end of the second generation the cats had developed degenerative diseases by mid-life. By the third generation, the cats eating the cooked foods had developed degenerative diseases very early in life and some were born blind and weak, and had a much shorter lifespan. Many of the third generation cats could not even produce offspring. Any cats making it to the fourth generation showed the diseases listed much earlier in life than previous generations. The diets lacking natural digestive enzymes created a succession of illnesses, while the enzyme-rich diet sustained complete health. Dr Pottinger conducted similar tests on mice. The results reinforced the cat study.

Could it be that the increase in autoimmune and neurological disorders in 'modern' society is the pattern that played out in the Pottinger study? There is a stark increase in the rise of autoimmune and degenerative illnesses in the general population, and children are showing intensely more developmental difficulties earlier and earlier in life. Our modern processed diet contains even more additives than Dr Pottinger used. The environmental toxin load is no better either. Although we do not have any evidence that taking enzymes over a long period of time is harmful, we do have ample evidence that having insufficient digestive enzymes can be very harmful.

Enzymes make it easier to get rid of waste, and take the stress off the pancreas, thyroid, pituitary gland, liver, immune system, metabolic enzymes, and so much more. So, how is the pancreas 'stressed'? When more processed and cooked foods are eaten, the pancreas must work harder to produce more enzymes. If you supply enzymes with food, the enzymes will digest the food and the pancreas is not signaled to produce more for this purpose.

Dr Howell combined the data from 12 other studies from eight different researchers working on different issues. He compared the weight of over 370 rats' pancreases with the type of diet they were on. The results showed that the rats consuming a diet of cooked food had pancreas weights three times greater than the rats caught from the wild eating a diet of raw foods. Like a muscle that enlarges when it works harder, the pancreas expanded in proportion to the work it was required to do. All the raw materials and energy devoted to digestion in the rats' bodies were resources that could have gone into other functions (Howell 1985). As a footnote, Dr Howell founded National Enzyme Company for enzyme research and development, which is one of the premier enzyme companies in the United States.

Enzymes help a variety of conditions

As you age, your body very slowly begins to lose the efficiency it once had in younger years along with a decrease in metabolism, including efficiently producing enzymes. As people age, they frequently have comments and complaints such as not being as able to eat spicy foods any more; not recovering as quickly from aches and pains; losing stamina; less energy; and oh, my aching joints! This decrease in overall health may be related to the decrease in enzyme production.

You would expect enzymes to be helpful in resolving enzyme deficiencies or something like pancreatic insufficiency (which they have proven to do very effectively: Graham 1977; Knill-Jones RP et al 1970; Griffin, Alderson, and Farndon 1989; Balakrishnan 1981). Other areas where enzymes excel may not be so obvious. A lot of people are concerned about cardiovascular disease, clogged arteries, heart attacks, and strokes. A contributing cause of the plaque is undigested fats and proteins that build up on the walls of the arteries. If plaque builds up enough, you get clogged arteries. Taking enzymes on an empty stomach will help clear out this plaque and buildup in arteries (e.g. Taussig and Nieper 1979). Enzymes eliminate the fat and protein deposits left in the body from years of past poor eating habits. Many cardiovascular problems are the result from not digesting fat properly. Current research is exploring the relationship between enzymes and reducing cholesterol to reduce heart attacks.

Before the 1960s, bromelain, a standard enzyme from pineapple that breaks down proteins, was commonly used for burn victims to keep the wounds cleaned out. The bromelain degraded and removed the dead tissue and infection (Hewitt et al 2000; Houck et al 1983). Starley et al (1999) treated pediatric burns with papaya (papain). Supplemental enzymes that break down proteins (proteases) are becoming more widespread for a number of uses. Proteases are of interest largely due to their anti-inflammatory properties, and their ability to support, enhance, and regulate the circulatory and immune systems (e.g. Gutfreund, Taussig and Morris 1978).

Circulation is also improved by enzymes because they can increase the surface area of the red blood cell, which allows it to carry more oxygen to all parts of the body. Enzymes selectively attack only certain cells, proteins, and other substances. Our red blood cells are selectively protected from enzyme degradation by identification mechanisms that protect them from breakdown.

Good circulation is needed to transport nutrients to the various parts of the body and move waste elements and toxins out. Although circulation naturally slows down as we get older, supplementing the diet with more enzymes will help. Good circulation is an important part of healing many diseases. Good circulation is necessary to deliver the nutrients to the right destinations.

Dr Max Wolf tested 347 patients with circulatory disorders. His findings showed 87 percent of the patients became completely free or almost free of any symptoms after taking enzymes. Dr Maehden treated 216 cases of various circulatory disorders with enzymes. Ninety-three percent were completely recovered or had significant improvement (Maehden 1978). Clinical proof such as this and many other studies (e. g. Worschhauser 1990) shows the more enzymes you take, the quicker your body can restore, repair, and strengthen itself.

Enzymes are effective in weight regulation. Depending on what you need, taking enzymes can result in weight gain or loss. Many obese people have very low levels of lipase in their body. Without sufficient lipase, the fats from food cannot be properly broken down and used. It may be stored instead. Taking lipase enzymes breaks down these fats. Also, when digestion is bad, the body craves more energy and nutrients. When food remains insufficiently broken-down and not absorbed, the nutrients within the food are not released and delivered to the cells to satisfy the need. Some people continue to eat more attempting to satisfy that nutrient hunger. When digestive enzymes lead to better nutrient release and absorption, the body is functioning better (more efficiently), metabolism is up, and this can lead to weight loss.

If the problem is being underweight, supplying enzymes can lead to better nutrient absorption, proper health, and thus, weight gain. Weight gain can also result because of reduced discomfort caused by indigestion, bloating, gas, and other problems, which may discourage someone from eating a balanced diet (Suarez et al 1999).

Enzymes tend to have the net effect of regulating weight into an appropriate range for the individual. Sansum's (1932) published results on weight from a number of cases at the Potter Metabolic Clinic show that weight regulation is a long-standing effect of taking enzymes. Of 197 individuals who were underweight, 91 percent improved by gaining weight with enzymes. Of 29 individuals who were overweight, 93 percent improved by losing weight. Of the 54 individuals who started at an appropriate normal weight, 100 percent (all of them) remained at a constant weight.

Real healing versus treating symptoms

Taking oral digestive protease enzymes is a nice alternative to the many prescription and over-the-counter drugs available. Drugs often focus on providing temporary relief from pain and other symptoms. A group of medications known as non-steroidal anti-inflammatories (NSAIDS) is readily available and prescribed frequently, including pain relievers such as acetaminophen, ibuprofen, and naproxen sodium. However, some medications may not actually decrease the inflammation or get rid of the problem. Rather they act by lessening the pain response. This means you just do not feel the pain of an injury, not that the injury has healed. Pain is a mechanism that lets us know that something is wrong in our body. It is much better to repair the damage than to just ignore the injury that is there.

Pain relief and rapid muscle and tissue healing is a great concern to those in sports medicine, including medical professionals and coaches, besides the athletes. NSAIDS provide temporary pain relief but can have dangerous effects on the health of athletes. These adverse effects include digestive problems, a decrease in overall performance, and the potential to make the injuries worse because the pain is masked. An athlete may feel better and use a muscle before it is strong enough. He risks re-injuring the same muscle or joint, sometimes even worse. Taking a pain reliever can be beneficial for our own comfort while the injury is healing; however, attention must be given to repairing the damage itself. Enzymes work to heal the injury, they don't just mask the resulting symptoms.

Sports injuries are often treated with protease enzymes because of their ability to reduce inflammation and speed the healing of bruises, swelling, and other injuries (e.g. Blonstein 1967; Bucci 1995; Masson 1995). A study conducted by Dr M.W. Kliene and his coworkers at the Sports Medicine Investigation Center in Grunwald, Germany researched the effectiveness of proteolytic enzymes on 100 athletes. The results favored the enzyme treated subjects. Seventy-six percent evaluated the success rate as 'good' with protease enzymes while only 14 percent of the placebo subjects reported a good success rate (Kliene *et al* 1990). The effectiveness of enzymes is so well understood in sports medicine that millions of enzyme capsules are sent to the Olympic training camp for both the German and Austrian teams at the start of the Olympics to aid the athletes in healing quickly.

Several studies showed that when enzymes were taken before surgery, the swelling from the injury left around seven days quicker, on average, than those taking a placebo; post-surgery recovery was also much more rapid (e.g. Duskova and Wald 1999). NSAIDS can actually slow recovery from injuries because they inhibit the body's natural physiological responses. Although studied extensively abroad, North America has not been as quick to recognize these conclusions, and adopt enzyme therapy.

The end of the line – Keeping the colon clean

When food is insufficiently digested for any reason, it can accumulate in the colon. While in the colon, undigested proteins can putrefy. Undigested starch, sugars, and carbohydrates can ferment, and fats can turn rancid. Regular bowel movements are necessary to sufficiently get rid of any toxins. When the rubbish is not removed quickly, constipation can develop, and toxins can build up and further damage the gastrointestinal tract. A gut full of waste is a prime breeding ground for unhelpful bacteria and other pathogens. These pathogens only add more toxins as they grow and excrete waste and by-products. If the gut is leaky at the same time, these toxins can cross over into the bloodstream and then you have a big mess. It is like a toilet overflowing back into your body. Yuck! It is best to keep the colon cleaned out as much as possible. A clean colon means less sludge in your bloodstream and less sludge in the intestinal tract.

A toxic colon can lead to dysbiosis, yeast overgrowth, constipation, and inflammation of the colon to the extent the person develops irritable bowel syndrome or a bowel disease. With all those toxins building up, the liver may become very overburdened. So now, add liver problems on top of your growing pile of maladies. Then the immune system jumps into action to try to restore order. Edelson as well as others identified that liver detoxification is a source of problems for many people with autism conditions (Edelson and Cantor 1998).

So, how do enzymes help the colon that far at the end of the digestive tract, especially if they might get reabsorbed at the end of the small intestine? First, if the food gets broken down properly in the stomach before it even gets to small intestine, the small intestine has less work to do and finishes the job quickly with the nutrients being properly absorbed. You will not have masses of insufficiently broken down food even reaching the colon. Then in the small intestine, enzymes can proactively help heal an injured gut and facilitate nutrient absorption and digestion. This is great! This helps ensure that the nutrients entering the mouth actually make it into the body instead of whizzing right on through and out the other end.

Next, the liver is not so overburdened because enzymes are very efficient at detoxification and keeping the bloodstream cleaned out from any gunk and large protein molecules that make it into the system.

Enzymes also help keep the gastrointestinal microorganisms in the proper balance. This is beneficial to the colon because the harmful bacteria cannot get out of control as easily. Enzymes making it to the colon (not every enzyme molecule is reabsorbed) will help to breakdown any waste building up there as well, cleaning out any crevices and curves, carrying off bacterial wastes, and other sanitary details. Should parts of the colon become inflamed, proteases will help keep the inflammation down and clear out any infection.

Some older literature may contain other theories and ideas of how enzymes work besides what is presented here. Since those writings, more has become known about enzyme function and behavior in living systems. Better instruments and assays are now available and there has been a tremendous surge in the area of medical biochemistry in the past few decades. This has led to improved and updated knowledge of how enzymes work and how they relate to other parts of living organisms.

Many of the medical situations given here are also very common in people with autism conditions. If enzymes are proven to work in all these areas, it is not surprising that enzymes would help people with autism. Maybe these were the reasons we were seeing so much success with the new enzymes. Maybe the formulation had tapped into this stash of underlying needs and was systematically restoring the body to health by several paths.

The elusive conclusive scientific studies

Why hasn't there been more research done in the area of enzymes and various neurological conditions? For one, several of these problems have been regarded as psychological in nature for a very long time. Since the concept and acceptance that many of the symptoms and behaviors seen in these conditions may have a very real and very legitimate medical basis, research in this area is increasing. Neurology itself is not an exact science or completely understood although much research continues in this area also.

We also have the current situation that most medical practitioners are not trained in the areas of nutrition, enzyme therapy, or other treatments that fall into the category of 'alternative medicine.'

Additionally, it is helpful to note that getting funding for studies that do not stand to bring a certain amount of financial gain back to the investors is hard to organize. Dietary supplements and therapies are often not patentable, and thus are not financially appealing. Enzymes are basic natural substances found in all living things. Companies find it very difficult to recoup monetary investments in this type of research, or in offering these not-as-profitable supplements. In the meantime, it is still very beneficial to look at studies using enzyme therapy for related conditions, such as infections, autoimmune disorders, and gastrointestinal disorders.

Getting a scientific study completed and published can be a very involved process. Good biological and nutritional research is very expensive and time consuming. Obviously, people suffering from a particular disease or condition may not have the luxury of waiting because time is going by and they need to do something now. However, nobody wants to be a 'guinea pig' for untested treatments either. Finding the right compromise in testing versus delaying availability of a treatment is a perennial debate in the medical, scientific, and regulatory communities. Let's say we eventually find a certain treatment is successful for 999 out of 1000 people. Sounds great, right? But if you happen to be that one person in a thousand, that is 100 percent 'failure' for you, and you may very well be understandably upset by any adverse reactions. But does that mean the other 999 people should be denied therapy until it is refined and tested more? How much testing is 'enough'? There will always be some risks.

A related issue is that some people will not accept the conclusions of a study even if done incredibly well. It could be they simply do not like the answer, or do not think the study was set up to address all the necessary variables. An example is the publication that some children are helped with some vitamin therapies. However, other studies done similarly failed to find the same conclusion. They found the opposite. Now what? Which do you go with? How is a person to decide if a piece of research is valid and done well?

Even if we have an outstanding new project conducted under prime conditions and achieve having it published in a very respected journal, does this give us the guarantee we are looking for? Is the issue closed, the final answer in? A study is still just one study under certain conditions looking at particular parameters. Studies are self-contained little worlds you need to balance in perspective with the broader spectrum of life. You have to evaluate it for yourself taking in all the surrounding considerations. Evaluate it in view of other research and other sources of information.

Often, as many questions arise from a study as they answer. The elusive, conclusive, all powerful scientific guarantee does not exist. Each person ultimately evaluates what is 'good enough' for them. Really, there is nothing that can prove and guarantee anything will work specifically for *you*. People are too unique, biology too complex, and science is not exact enough for that.

CHAPTER 9

What to Expect When Starting Enzymes

Drat! A glitch! Dana posted an update three weeks after she reintroduced wheat gluten to her son. She had been giving one piece of bread, once a day for three weeks with enzymes. At first, her son was doing fine with it, but now he was sliding back into negative behavior. Not only that...he developed a rash and diarrhea. Now what?! Was there an allergy to the enzyme formulation? Were the benefits of the enzymes wearing off? Regression?

Dana had been eliminating casein and gluten as well as many other foods for a year and a half at this point with three of her four children. The oldest exhibited ADD, the second one was diagnosed very low-functioning autism, the third child showed PDD/language delay, and the youngest was neuro-typical but had some food issues. Dana had been eliminating foods until she was down to five foods (I kid you not, F-I-V-E) that did not provoke a negative reaction. Dana said that with the new enzymes she was still able to give many other foods, just not gluten, so she considered these enzymes very successful. With the enzymes, she was now able to reintroduce enough different foods that she could buy some of the packaged foods that contained ingredients such as honey and potatoes. She was now able to go *on* a casein-free, gluten-free diet. Imagine that! Someone who considered being on the casein-free, gluten-free diet an expanded menu.

But it was disappointing to hear that gluten was not working out for her when others were doing quite well. And this bothered me. Why wasn't it working? It didn't make sense for enzymes to just 'stop working.' In the end, Dana investigated the rash and discovered it strongly resembled the type of rash seen in some cases of celiac disease. About this time, two other people who knew they had celiac disease found they too could not eat gluten with Peptizyde. In fact, not only could they not eat gluten, but when they used Peptizyde, the negative reaction appeared to be even worse than before. There was no problem taking Zyme Prime (a broad-spectrum product). Now this truly didn't make sense…at first.

I investigated and others investigated the situation of celiac disease. Two more people reported very negative reactions with Peptizyde and gluten, but not the Zyme Prime. One knew her child was celiac, but the other one said no. At least we had a trend going even if it was only a few people.

Over the next few weeks, three more cases turned up. In each case, the person having celiac disease had a negative reaction when they consumed gluten and took Peptizyde. The advice was now that Peptizyde was not allowing a person with celiac to eat gluten. However, it did allow most of those with celiac to have dairy again as well as other foods. After a few more months, I uncovered enough research to put together a reasonable explanation for what we saw.

Celiac disease

At this time, enzymes are not proven to allow a person with celiac to return to eating gluten. Celiac is a serious autoimmune disease/ condition with a genetic basis that often goes undiagnosed. So, if someone sees adverse reactions gluten plus enzymes, especially the strong proteases, they are advised to consider celiac disease as a possibility.

There is a certain structure in the gluten or gliadin substance that the small intestine sees as toxic in celiac individuals. This is a different situation from a peptide problem caused by insufficiently broken-down molecules, or leaky gut. Those peptides have a certain structure that might attach to specific neuro-receptors in the brain or gut, or elicit an immune system response. Celiac reactions are triggered by specific tiny peptides that attach to different receptors in the small intestine. It is also different than IgG-mediated food intolerance. Only celiacs appear to have intestines that respond in this way.

When the peptides get to the small intestine, the celiac's body registers these peptides as 'the enemy.' A non-celiac's body would just see the peptide as being from gluten and let it pass. Once the celiac's body detects 'the enemy,' the immune system kicks into gear and attempts to deal with it. One theory is that the proteases might very efficiently break all the larger gluten pieces down into tiny minuscule parts, and so create many more little parts to aggravate the villi instead of a few larger parts in people with celiac. No two celiacs are alike in their dietary tolerances for gluten – some are very sensitive, some can tolerate a little at a time, some cannot eat oats, spelt, or kamut, whereas others can.

I found some research indicating the part of the gluten or gliadin that is harmful to celiacs may not be the protein itself but rather a carbohydrate portion of the gluten or gliadin (Phelan *et al* 1977, 1974; McCarthy 1976; Hekkens 1963). When the proteases break up the gluten, although this destroys any problematic intermediate peptides, you may still have many little carbohydrate pieces around that may upset the celiac intestines. Peptizyde does not contain enzymes for carbohydrates whereas Zyme Prime has a great deal. Zyme Prime has much fewer proteases than Peptizyde, too.

The research showed that by taking amylase and glucoamylase to breakdown the carbohydrate portion of the gluten as well, the test subjects were able to consume the gluten without problem. Additional studies supported this showing when the carbohydrate portion of gliadin was manipulated the problems ceased (Phelan *et al* 1978, Stevens *et al* 1978, Anderson *et al* 1981; Gottschall 1994). Other research identified certain digestive enzymes that could be the basis of an enzyme product to help celiacs with gluten (Hausch *et al* 2002).

The practicality of this working in everyday real life has not been worked out yet, nor is there a Celiac-zyme. It does, however, explain why people with celiac do so much better with Zyme Prime.

Perhaps because it contains a substantial amount of amylase and gluco-amylase with lower levels of proteases.

So, the take-home message is if you have celiac, do not consume gluten even with enzymes targeted for gluten at this point. If you suspect any gluten, even trace amounts, you may want to take something containing starch digesting enzymes for the celiac issue.

Since this discovery, many people with celiac are doing very well on enzymes for other foods, just not gluten. For more information on celiac disease and photos of the related rash, see *www.celiac.org.*

Adjustments, side-effects, troubleshooting

There were several other reports of mild side-effects. Most ended by the third week, were considered quite manageable, and not too disruptive. This establishes the guideline that enzymes should be tried for at least three weeks to allow for common adjustments by the body. Probably one additional week should be added for good measure in case you are just starting to see positive effects in the third week.

Researching into the matter, we learned that many of these reactions are not serious at all and quite common when starting any type of digestive enzyme. Many of the adverse reactions are really the body cleaning out and healing itself. Some side-effects are also the same as those seen when starting a restrictive diet. Most side-effects result from the body adjusting to:

- processing more food and liquid
- gut healing, cleaning out inflammation and waste
- die-off from yeast, bacteria, or other pathogens in the gut
- withdrawal effects from 'addictive' food compounds

Hyperactivity, irritability, and withdrawal or allergy-like symptoms can actually be indications the enzymes are working and the body is adjusting.

Hyperactivity
Hyperactivity or an increase in self-stimulatory behavior (stimming) seems to be a very common early adjustment effect with enzyme use, although by no means does every person react this way. Several reasons

why a person may be more hyperactive with enzymes initially and some suggestions for dealing with it are given below.

1. Hyperactivity may be a 'withdrawal' reaction from gluten, casein, sugar, carbohydrates, or other 'addictive' substances. This can happen even though you have been on a restrictive diet for some time. Enzymes will be breaking down hidden or unknown sources of problematic foods. Taking enzymes may be removing the possible sedative effect of certain compounds. You may want to decrease the enzyme dose to lessen any excessively unpleasant symptoms. Hyperness as well as other undesirable behaviors is also a very common behavior when starting some types of restrictive diets, including Feingold program, casein-free, gluten-free diet, or yeast control diets. So it is reasonable to see the same reactions with enzymes because, in many cases, they are accomplishing the same thing as food eliminations.

2. The person taking enzymes may be trying to adjust to an increase in awareness or sensory input. When there is significant improvement in many areas, there may also be an increase in stimming, hyperactivity, anxiety, and a bit of sound sensitivity. Realize that as the person becomes more aware, socially and otherwise, he has a lot more sensory input to process and interpret. At first, this can be overwhelming and frustrating. It could be that he resorts to the stimming and hyperactivity as a protective behavior to try to calm or organize his system. This appears to be very common or typical.

I find when I am overwhelmed by a lot of sensory input, I tend to just start shutting parts out. Someone watching me might say I am withdrawing into my own little world and isolating. I am trying to prune out the overwhelming amount of sensory input to a level I can manage. Then I regroup, reorganize, and emerge once again. This is similar to an army that goes into battle with a good plan and resources. Then communication lines get cut, bridges break, and eventually you get total confusion. So the troops are recalled, the situation reassessed, new plans are made, and then they go forth once again.

The challenge comes when my level of 'a lot of sensory input' is not what other people consider a high level. Such as going to a fast-food restaurant with my Fortunately-Very-Handsome-Husband who is chattering away trying to 'help' me order while I am trying to read the menu, trying to answer a string of questions like 'would you like fries with that,' trying to manage being around so many other people, and think all at the same time. I need at least one of those things to go away. Two would be better. From the outside it may look like I am just standing there doing nothing or in a daze, but inside I am under a sensory assault.

3. Enzymes are known to help keep yeast and bacteria in check. When digestion starts to work better, you have better nutrient uptake and less malabsorption. This leaves less food for adverse yeast or bacteria to feed on, and they die off. In addition, enzymes can break down the yeast, bacteria, and pathogens directly. As the microbes die, they release neuro-toxins and substances that can produce allergy-like symptoms. This die-off reaction is common with other yeast and bacteria treatments. You may see more frequent bowel movements, increased irritability, increased or decreased appetite, flu-like symptoms, hyperactivity, and behaviors that are more compulsive. Usually this readjustment period ends by the third or fourth week if enzymes are taken regularly. The enzymes may help remove many of the toxins produced by these little critters. Some people find that giving activated charcoal is very effective in reducing these die-off symptoms. The charcoal readily binds the toxins carrying them out of the body. Drinking more water helps flush the toxins out as well.

4. Better digestion and overall improvement in health caused by the enzymes may unmask a nutrient deficiency. A common deficiency is magnesium. Symptoms of magnesium deficiency are hyperactivity, anxiety, and muscle spasms. Magnesium is recommended for treating hyperness in many conditions besides autism spectrum or attention deficit conditions. Adding a soluble form of magnesium may be very helpful with no known toxic levels, and is inexpensive to give. Other supplements known to help with hyperactivity are calcium, zinc, folic acid, and molybdenum. Taking digestive enzymes along with some of these supplements may help the body to make better use of these minerals. (see Chapter 16 on Magnesium and Neurology)

5. Many parents report that adding Zyme Prime with the Peptizyde has decreased or eliminated the hyperactivity. No one really knows why for certain. Not only did the hyperactivity decrease, but the child became very pleasant – the Happy Child Effect. Perhaps the glucose regulating ability of the carbohydrate enzymes in Zyme Prime have an effect. Or maybe the Zyme Prime makes more nutrients available from food, such as molybdenum, sulfur and magnesium, which are needed in the processing of phenols and other chemicals. More magnesium also helps with hyperness in general. One theory is that an all-purpose enzyme product that is not as high in proteases or meals that are not as high in protein may favor an increase in serotonin levels. An increase in serotonin helps with calming in most people.

6. There may be a problem processing phenolic compounds, a certain chemical that occurs naturally in all foods. When phenols cannot be processed out, they build up in the body exerting a toxic effect. Hyperness is a common symptom of this. Many people with autism and attention deficit conditions do not process sulfur adequately. Poor sulfur metabolism leads to problems with processing out compounds such as phenols. Epsom salts (magnesium sulfate) are very helpful for many people because they supply both magnesium and sulfur in usable forms. Many parents noting hyperness were able to resolve the situation successfully by providing Epsom salts on the skin. (see Chapter 17 on Sulfur, Epsom salts, and Phenols)

In addition, No-Fenol from Houston Nutraceuticals is a new enzyme product showing great results for alleviating the negative reactions seen when sensitive individuals consume highly phenolic foods. Fruit-derived enzymes, such as bromelain, papain, or actinidin, may aggravate this situation as well, so looking for products without these particular enzymes may be helpful if you are very sensitive to phenols. Enzymedica is one company that makes several types of enzyme products without fruit-derived enzymes. A version of Peptizyde, AFP, does not contain fruit-derived proteases either.

Also, consider removing any artificial additives and any chemicals added to foods. These can cause substantial health problems which manifest as hyperness, behavior problems, and physical symptoms.

7. There is a principle that too much of something can be as problematic as too little of something. Reducing other supplements or discontinuing them may help with hyperactivity or other adverse reactions. One mom put it this way: When you go to the gas station and you need 10 gallons of gas to top off your tank, putting in 11 gallons is going to cause a big mess. This describes well a concept and pattern we noted in the enzyme group. Sometimes nutritional supplements that helped initially seemed to become more problematic after starting enzymes. Even after foods were carefully added or eliminated with enzymes regularly, their child continued to show reactive symptoms such as red ears, poor sleep, and hyperactivity. Increasingly, this became associated with some of the supplements. By withdrawing or reducing the supplements, the physical markings and especially the hyperactivity stopped. This seemed to be particularly common with any high B-vitamin supplements. Some of the oils, such as flax seed oil, fish oil, and borage seed oil, were also identified as culprits causing hyperness or aggression.

Research confirms that nutrients in too high a quantity, or out of balance can have a negative impact on health. Chandra (1986) finds that certain nutrients in excess can suppress the immune system. Or too much of one nutrient causes a deficiency in another one, which results in new or different problems.

Let's say we have a child with a nutrient capacity of 10 to 15 'gallons' or 'units' for proper functioning. Over 15 units creates a problem and under 10 units creates a problem. Now, if the child's body can only produce about five units of Nutrient X, he is going to have problems. He needs to stay between 10 and 15 units of Nutrient X.

Window of Optimum Health = 10 to15 units of Nutrient X
Body produces 5 units – too low, regression

So now we give a supplement that provides 6 units of Nutrient X with each dose. In the beginning, the additional 6 units plus his regular 5 units keeps the body at 11 units, which keeps his total within the range for optimum health. And we are thrilled with how 'effective' this supplement is.

> Body produces 5 units + 6 units from supplement = 11 units
> see improvement, Yay!

When we start enzymes, the enzymes may break down food such that more of Nutrient X is made available to the body. And since overall health may be improved, more of Nutrient X may be absorbed and used naturally. This starts to increase the body's natural amount from 5 to 6 units, then to 7, then to 8, and so on headed right up to proper health and functioning.

> Body produces 6 units + 6 units from supplement = 12 units
> see improvement, Yay!

> Body produces 7 units + 6 units from supplement = 13 units
> see improvement, Yay!

> Body produces 8 units + 6 units from supplement = 14 units
> see improvement, Yay!

> Body produces 9 units + 6 units from supplement = 15 units
> see improvement, Yay!

At first, enzymes plus the supplement are both looking good...because although the total amount of Nutrient X is rising overall due to better nutrient availability and absorption, the amount is still in the healthy range. However, it may get to the point where the body plus food plus enzymes plus health reaches 10 units naturally on its own. Now the additional 6 units provided by the supplement = 16, and perhaps moving on to 17 or 18 units. We are now over the limit, and we have 'regression' or a mess again.

> Body produces 10 units + 6 units from supplement = 16 units
> too much, regression

> Body produces 11 units + 6 units from supplement = 17 units
> too much, regression

> Body produces 12 units + 6 units from supplement = 18 units
> too much, regression

Then the parent hears about this potential problem and takes out the supplement. This brings the total back into the acceptable range and the child improves once again.

Body produces 12 units + no supplement = 12 units
healthy range, Yay (again)!

This in no way means the supplement was not any good or was a mistake to give in the first place. This just means that it is not necessary anymore, and for a very good reason – the child is in better health. Ending handfuls of supplements would be a reasonable goal of using enzymes. Eventually, it may be possible to even end the enzymes themselves, should you want to. Several people report needing less enzymes over time, not more, to maintain the same level of physiological health.

Overall, most parents with a child showing hyperness when starting enzymes saw that this resolved itself (went away) over time, or as they pursued one of the above recommendations. This is another reason to allow a trial period of three to four weeks when beginning the use of enzymes. If the hyperness is at an unacceptable level, then decrease the dose and build up more gradually. You can start as low as one-eighth of a capsule or even a few sprinkles of enzyme per meal or snack. This allows the body to recover and adjust at a more tolerable pace. Having the person stressed, intensely uncomfortable, or 'toughing it out' is not necessarily the fastest way to improvement.

Special note on enzymes and medications
There is usually no chemical interaction between enzymes and most medications. Always be sure to ask about any potential interaction if you will be taking any medication and enzymes at the same time.

Some families were considering medications, such as an antidepressant or stimulant, but once they started enzymes, their child improved to such a degree that the medication was no longer considered necessary. I no longer have the continuous muscle tension resulting from migraine pain, so I no longer need to take baclofen or other measures all day long to counter this.

One possible interaction is that the enzymes may be breaking down the medication more thoroughly, or healing the gut so that the medication is now being absorbed more completely. This means that the person may now actually be absorbing or metabolizing a slightly higher dose of the medicine than before; their body is now reacting to slightly more medicine. If you are giving a particular medication, please be aware of any possible effects from the body responding to a slightly higher dose than before. A few parents have noted side-effects when starting enzymes, but they were side-effects from a medication now being absorbed more thoroughly. In one situation, the medication had a side-effect of drowsiness. This is not a typical effect of enzymes. Decreasing the medication ended the effects.

There is one caution with time-released medications and the enzyme cellulase. Some time-released medicines use cellulose to slow down the digestion and release of the product. Not all do. If you are taking a time-released medication and want to use enzymes, please consult with your doctor or pharmacist.

In addition, several people have been able to reduce or discontinue medications they were on after starting enzymes. My family was able to reduce the amount of both our medicines after taking enzymes for two months. I would say it is probably because the benefits from enzymes decrease the need for taking as much medication to begin with, but we also might be absorbing and metabolizing it better.

Regression
Some parents see signs of 'regression' when they begin enzymes. This may be due to the enzymes clearing out peptides (maybe internally produced ones), sugars, or other substances from previously unknown sources, so you see withdrawal symptoms similar to what you might have seen when beginning a restrictive diet. Any of the above reasons given for hyperactivity (die-off, sensory adjustment, adjustments in other supplements) may also lead to temporary regression.

Thirst
Many people notice increased thirst when they start enzymes. This is perfectly natural. The body is metabolizing more food, so more water

is needed to process the food. It is most likely a sign that the body is functioning better and absorbing more nutrients. Just give more water.

Irregular bowels

Having looser stools for the first couple of days is very normal with any enzymes. This may be because the body is experiencing more water as the enzymes break bonds – one molecule of water is produced for each broken bond. More food and liquid is being processed as well. The loose stools usually regulate within a few days, but may last up to a week or two. If it persists longer than that, reducing the amount of enzymes you give may help. Or you should probably discontinue the enzymes and seek the advice of your doctor, the manufacturer, or others using enzymes.

Wetting

This tends to be one of the most annoying adjustment effects overall. A very small minority of families see temporary night-wetting, along with more frequent urination when starting enzymes. This usually disappears within the first week but may persist for two to three weeks. In a couple cases wetting continued up to the fifth week. The body adjusts with a little time. Each molecular bond broken by the enzymes produces one molecule of water. The more food that is broken down, the more water is created. Proteases are making more amino acids, carbohydrases are making glucose, and their increased presence may act as a rehydration mechanism. If enzymes are breaking bonds on waste and toxins, there will be more water.

Stomach ache

Sometimes after starting a strong protease enzyme product, the person may have a stomach ache or pain when eating. The general recommendation is to stop the proteases for around four or five days, and then try again. Usually this resolves the problem. Although proteases such as bromelain and papain have been used extensively to heal gastric ulcers, proteases may be irritating to the gut initially, particularly if the gut is very damaged or inflamed. The proteases will be cleaning out wounds and inflammation, taking toxins and debris away. This may leave exposed healthy, yet sensitive tissue. Stopping the proteases for a few days allows these cleaned out, but exposed areas, time to heal again. The injured areas may not be as sensitive when proteases are later resumed.

Another strategy would be to start with an enzyme product that is low in proteases (such as a general all-purpose product), which will promote gut healing, and then adding in the stronger protease product after a week or more. If you wait the four to five days, resume the proteases and then see stomach aches again, this may be a symptom of celiac disease or something that requires different measures. You may also try reducing the dose, or stop the strong proteases altogether. There may be more substantial gut injury which needs longer to heal before stronger proteases can be used.

Reaction to formulation, including mold

Some people may react negatively to something in a formulation itself, which is possible with any product. Some people that react to rice or beets are able to tolerate products containing rice bran or beetroot powder because the enzymes break down these components as well as other foods. Another possibility is that a person will react to a particular mixture as a whole and no specific item is identified as the offender.

Often people ask if there is mold in microbial or fungal-derived enzymes. The answer is no, there is not. Microbial or fungal enzymes are derived by isolating microbial strains that produce the desired enzyme. This technique, called fermentation, has been around for over 3000 years. Once fermentation is completed, the microorganisms are destroyed. The enzymes are further processed and refined for use. All plant and fungal enzyme products, regardless of manufacturer, are primarily derived from papaya, pineapple, kiwi, or *Aspergillus oryzyae* or *Aspergillus niger* microorganisms. Microbial enzymes are secreted by the fungi, which are cultivated in large incubators. The enzymes are separated and purified away from the rest of the fungus through a process involving eight to 12 different protocols. As a result, the final enzyme preparation does not include any part of the fungal body (usually the allergenic part).

This situation is the same as with penicillin and bread mold. Penicillin is purified from bread mold, but when one gets a shot of penicillin, they are not getting a shot of mold! In the same way that some are allergic to penicillin, some can be allergic to plant or fungal enzymes, but this may be totally unrelated to any reactions to molds. You can develop a sensitivity to any protein. Because enzymes are proteins, there is a very slight chance that someone may react negatively to any enzyme itself. You may want to be cautious though if you have a true allergy to molds. In actuality, many people who are reactive to molds have no trouble with plant- or fungal-derived enzymes. Animal-derived enzymes are an option if you cannot tolerate microbial or plant enzymes for any reason.

A few people so far have had problems with the enzymes papain (from papaya) and bromelain (from pineapple) which are in many protease containing products. Papain and bromelain are cross-listed for reactivity along with kiwi, some seeds, and pollens. This may be related to a true allergy. So, if you have problems with any one of these, also be cautious with other products containing any of these other ingredients. Papain is also listed as cross-reactive to latex allergy.

There have been no reports among the hundreds in the enzyme group since it began May 2001 until this writing where someone 'became' reactive or intolerant of an enzyme formulation itself over time. Several times a child would show regression, but in every case it was later found to be due to some other cause such as yeast infection, a cold, traveling, new therapy, or other supplements. Or regression may happen if enzymes are used with certain foods that they are not adequately forumulated to break down.

In the first four months, there was some speculation that an increase in hyperactivity or aggression may be due to the L-glutamine that was present in the initial formulation of Peptizyde. L-glutamine helps with gut healing, and this was the reason it was added to the product. However, like many amino acids, it also has excitatory properties – meaning that it stimulates some central nervous system pathways. L-glutamine was suspected of causing some of the adverse reactions and was removed in September 2001. The current blend in Peptizyde does not contain this ingredient. There was a dramatic drop in the reports of aggression and negative hyperactivity after this change in the enzyme blend. Several parents who used both formulations said their children were much calmer on the formulation without L-glutamine, if they saw any difference at all. This illustrates that there may be some ingredient in any particular product causing an adverse reaction in an individual.

Hypoglycemia

This is not a true adjustment effect or side-effect of enzymes, but it may be the cause of some of the negative behavior seen. Hypoglycemia is commonly overlooked, but several parents have found that making adjustments for it evens out the moodiness they see in their children. When you are very sensitive to changes and have a neurological feedback system that may not be in the best working order, changes in blood sugar may produce a big effect.

Hypoglycemia means 'low blood sugar.' The glucose (a form of sugar) in the blood gets low. Glucose is the food for the brain, so when the glucose supply drops too low, the brain goes hungry and things happen. Giving protein along with carbohydrates tends to even out the reaction. So, cheese with crackers, peanut butter with apples, or some combination like that works well. Having smaller and more frequent meals also helps. A practical schedule is three small meals and three snacks each day. This means you eat within every three hours or so – a good target goal. You may have a sugar intolerance as well, providing another reason why adding a broad-spectrum enzyme product helps. (see Hypoglycemia in Chapter 19)

Adjusting to change - Hyperactivity/stimming perspective from an autism spectrum adult

This is a perspective on how hyperactivity, stimming, and regression may be related to improvements, and a sign the enzymes are working, as written by an autism spectrum adult. The view is that at a certain point, some actions may be just nervous habits, like everyone has. But if the habit is not one of the commonly accepted nervous habits, such as tapping your foot or flicking a pencil back and forth, then it is seen as some weird problem that must be stopped. Depending on what the behavior is, and how intense or frequent it is, you might consider just helping the person find a socially acceptable alternative to do that also lets off nervous tension. The term 'stimming' or 'stim' refers to any of a number of repetitive behaviors commonly seen such as rocking or flicking your fingers in front of the eyes. It means a self-stimulatory action although it can be used for calming and to help the person focus.

> Think of when adults get really involved in solving a problem. Okay, so here are these adults. They sit there and they are staring at the statistics, numbers, problems, or whatever. Maybe they are trying to figure out something they do not understand such as why some kids with autism spectrum conditions show an increase in hyperactivity and stimming while taking enzymes.
>
> But I see them sitting there, staring at the statistics, chewing gum or their lips, picking their teeth, tapping or twiddling their pencils, tapping their feet, pacing across the floor, or (this is my favorite one) bending all the paperclips a certain way and then lining them up on their desks. Do you get my point here? What is so bad about kids increasing their stimming? These researchers are trying to figure out something which does not make sense. They are trying to completely *focus* their mental energy and apply themselves 100 percent to figure this out, they need a release of all the confusion and nervous energy, so they STIM.
>
> Now let's consider your own child. For the first seven years of his life (or whatever age he or she is), you have seen whatever behaviors or difficulties caused you to pursue a diagnosis for the differences. Maybe he has walked around, no make that 'floated' through, with his nose in a book, seemingly oblivious to his family and surroundings. He LOVES his Legos (or whatever), to the exclusion of everything else. His brother or sister is just an annoyance.
>
> But now he is taking enzymes and is back 'with you.' He no longer floats through the house in oblivion. He is aware. He plays with his brother now, he is now interested in the pictures his sister draws. Maybe even his teacher has noticed big changes.

Based on my own personal experience, I would say that the decrease in hyperness and/or stimming etc., which may or may not be present when you start enzymes, is because the child basically feels terrible for a few days or weeks. Things are so darn confusing, nothing makes sense anymore, no energy to try to figure out why, so just try to sit back and be an observer and try to figure out what is going on without doing something stupid to attract too much attention to yourself. This may be expressed as withdrawal or regression.

Now, things look better. People are actually *interesting*, even more so than Legos! Certain concepts and things are not so confusing any more. Well, then I think your child would be doing exactly what the researchers are doing, I will repeat from what I wrote above, substituting 'autism spectrum children' for the word 'researchers.' These [autism spectrum children] are trying to figure out something which does not make sense, they are trying to completely *focus* their mental energy and apply themselves 100 percent to figure this out, they need a release of all the confusion and nervous energy, so they STIM.

I would also imagine that not only is it new and different and probably nice for them, but it is also very overwhelming because it is all so new and different. So they revert to what they know best when things get overwhelming, they stim, and it helps them settle down so they can learn about life again in this new presentation.

So why is stimming such a bad thing here? There may be potential peer pressure or teasing, which will encourage the child to learn to modify stims to be more socially appropriate. Because everyone stims. You just have to figure out which ones are acceptable, then the kids will fit in better by doing the same socially acceptable stimming behaviors as everyone else is doing, and everything will be just fine.

How can I tell if this reaction is good?

So, if your child is uncharacteristically bouncing around the house, is this hyperactivity good awareness or a negative reaction? Most of the time, positive behaviors occur at the same time as the adjustment effects listed. As long as you are seeing positive results along with any negatives, this usually means the adverse symptoms will resolve themselves shortly and the person is responding well to enzymes.

If you only see negative symptoms and no positive responses at all past the first three weeks, discontinue enzymes and/or consult your doctor, the manufacturer, or other people using enzymes. Often these sources can review your situation and lend suggestions on what you might be able to adjust or do.

If you or your child shows only positive reactions and no negatives at all, then count your blessings. This is common, too.

Like everything else, enzymes may not help every single person. If you know you have a yeast or bacteria overgrowth, you can expect to see more reactions from die-off to get through and positive improvements may take a little longer to see. Yeast or bacteria flare-ups tend to mask the benefits the enzymes are providing. Having excessive environmental toxin accumulation may also cloud improvements with enzymes.

One situation comes up maybe three percent of the time. No reaction is seen one way or the other. Nothing. No change at all. We do not know exactly what to attribute this to. A few people who saw no results with enzymes went on to detoxification therapy and saw some nice improvements with that avenue. Another possibility is that there may still be a food or chemical that needs to be removed and it is masking other improvements. A different enzyme formulation may give better results as well.

At this point, if no reaction at all is seen by the third or fourth week, then it is up to the individual whether to continue with enzymes or not. Older children and adults may need a little more time to see improvements. The enzymes might be working well behind the scenes and we just cannot tell. Continuing enzymes while pursuing other avenues has many advantages. The gut may be in better condition, nutrition improved, and the body better supported while implementing other measures to improve health. However, it is also hard to justify spending time and money on an avenue that is not producing visible results.

CHAPTER 10

Nutrient Deficiencies and Malabsorption
– Just Passing Through

After three months with enzymes, Cindy shared this update as a comprehensive evaluation of her son's progress. She was one of the first people to go from strict food elimination to enzymes plus most all foods again. Cindy is in the unique position of being a former special education instructor trained at the Master's level, so her view of looking at these things is slightly different from being a mom. You can decide for yourself the extent of any 'crash' over time.

Three-month Update, July 22 2001
by Cindy Kelley

My son is eight years old and has Asperger's Syndrome. He was on the casein-free, gluten-free diet for one year. I also had to remove eggs, corn, soy, nuts, tomatoes, mustard, chocolate, yeast, sorghum flour, artificial colorings, most flavorings, all but three fruits, most vegetables, and coconut butter. He was restricted to low-sugar, low-fruit, and no vinegar due to a yeast problem. Just before adding enzymes, I was planning to remove beef and white grapes due to apparent sensitivities developing. Any diet infraction consistently resulted in a bright red bottom, red ears, atypical behavior, and wetting. Usually the behavior pattern was a day of lethargy and unresponsiveness followed by two days of unexplained anger, aggression, and sound intolerance.

Three months ago, on April 21, my son had his first dose of Peptizyde and HN-Zyme Prime. He ate a complete restaurant meal of gluten and casein filled food. He had no adverse reaction. I took all the information about the enzymes to my son's pediatrician. She sent a request to the school that the enzymes be given with any diet infractions. We continued to give him the enzymes with each meal and allowed more planned infractions. We saw none of the previous negative physical or behavioral reactions – including no aggression. In fact, we started seeing an increase in many areas of development. At a school conference, my son's teacher and speech therapist insisted they were seeing significant changes that they attributed to the enzymes started two weeks earlier. At four weeks, with a much expanded diet, my son's teacher called me with more improvements in behavior. She saw a new enthusiasm for learning, a large increase in problem-solving skills, significantly more peer interaction, and an overall happier child.

Within a few more weeks, my son was completely off all previous dietary restrictions. The only side effect was wetting, which has since resolved. We now see the following behaviors on a consistent, daily basis:

1. Eye Contact: Increase of at least 500% since taking the enzymes. Eye contact now could be categorized as 'typical.'

2. Sound Tolerance: Previously could not tolerate brother's voice about 90% of the time and either had to leave the room or would scream at his brother. He now responds typically to his brother's sometimes excessive talking.

3. Initiating Conversation: Pre-enzyme, conversations were almost exclusively about his narrow field of interest (Legos, computer games). Now initiates conversations about a range of topics. Asks questions that reach outside him, such as, 'So, Mom, did anything wonderful happen to you today?' and 'What did you talk about while I was gone?'

4. Affection: Gives more spontaneous hugs; says, 'I love you' more. Overall, he is significantly more affectionate with parents and grandparents.

5. Empathy: Shows genuine concern for brother, which was not seen before enzymes. Yesterday and again today, my older son asked what was wrong with his younger brother when he looked upset. Today, I hung a picture on the refrigerator the younger boy made. My older son has *never* even noticed any artwork by his brother. Right after getting home from school, he saw the picture and said, 'Wow! He made a great picture! I need to congratulate him!'

6. Obsessiveness: No longer obsessed with Legos. Pre-enzyme, would always have a Lego flyer, book, or instruction manual in his hand. Legos are now a favorite hobby, but not an exclusive obsession. Now asks to do a variety of activities.

7. Self-stimulatory Behavior: Hand-flapping decreased significantly, especially in brightly lit stores. Since adding back zinc the last few days, stimming has decreased even more. He shows an overall decrease of at least 50%.

8. Self-injurious Behavior: Biting and hitting self when angry or stressed has decreased about 75% after enzymes. Since adding back zinc, these behaviors disappeared.

9. Transitioning: Previously, changing activities would usually result in back-talking or tantruming. Since enzymes, these behaviors have decreased about 80%. He now back-talks in a more typical way, not as an automatic and exaggerated response to all changes and requests. He is significantly better able to handle changes in daily routine, such as wearing mismatched pajamas or brushing teeth in a different bathroom.

10. Desire for Physical Activity: Pre-enzyme, we had to force our son to go outside to play. He is now compliant and even enthusiastic at the suggestion at least half the time.

11. Physically: His skin is less pale. Relatives comment he looks healthier. Circles under his eyes have lessened and his face looks more filled out. Stools are firmer, no longer light colored.

12. Interaction with Brother: By far, the most notable change. He plays with his three and a half-year-old brother cheerfully all the time now. This is one of my greatest joys. Before the

enzymes, he usually preferred his brother to be far away from him. I could always pick up on a new food sensitivity by watching for his response to his brother's voice. Pre-enzyme he would rarely ever play with his brother. Now he teaches him games, asks his opinion about topics in books, and wants to watch videos together. He points out observations and explains his ideas. He tells his brother jokes and giggles at his brother's jokes. Fights in a typical sibling way, instead of screaming 'Go away from me.' After enzymes, he asked to sleep in his brother's room for the first time. He now tolerates his brother's tantrums and misbehavior exceptionally well.

13. Mood: He is happy ALL THE TIME. Pre-enzyme, he was in a grumpy mood almost all of the time. He was often just hard to get along with. If he had a food infraction or developed a new sensitivity, he was angry, aggressive, and distant for up to three days. Getting him to do tasks was a chore. Now he is doing things willingly (well, like a typical child). He just generally seems to feel better – a significant increase in overall happiness.

14. Awareness: Significantly more aware of the actions of others. Since enzymes, he eavesdrops on our conversations. No longer appears detached. His humor is now more sophisticated. He recognizes the subtleties of humorous situations.

15. Problem Solving: Since starting enzymes, problem solving is no longer a weakness. He is able to reason better and does not give up readily. He is more aware of consequences and acts accordingly. Memory is better. Now remembers requests and general information better.

16. Speech: Pre-enzymes, his voice was unusually high-pitched. His voice is now significantly and more appropriately lower. Pre-enzymes, his voice was more flat. Now has a natural rise and fall within each sentence. Pre-enzymes, his voice was slow and measured. Now has a faster, natural rate of speech.

17. School: My son's teacher told me that since my son started the enzymes a few weeks ago, she has seen the following:

My son now looks forward to lunch, even tries to run to the lunchroom. Previously, he lacked any enthusiasm and had to be prodded to prepare for lunch. He is now cracking jokes often. He is talking out in class because he is so exhuberant about learning. He is more impulsive, but in a positive way, because he is so excited about participating. He is initiating conversations more. He is more assertive in general and is reacting to problem-solving situations in the classroom with enthusiasm. He shows more true emotion. The teacher said she feels she is seeing something like the movie *The Awakenings* where the patients come to life after taking a new medication.

Cindy's story is located at *www.enzymestory.com*

Cindy was like many parents who begin eliminating foods. What started out as a food intolerance to casein and gluten then became an intolerance to a whole list of foods. She continued to remove food after food after food until there was very little left. Was her son really not going to be able to eat all these foods for the rest of his life? Was he truly intolerant or 'allergic' to 30 to 63 different things? This is very representative of what happens when there is a highly permeable or leaky gut. When the intestinal lining is damaged, anything you eat might be insufficiently digested, and thus produce negative reactions. After a relatively short time on enzymes, just weeks, Cindy's son was back to eating most everything imaginable, and his behavior and health was far better than it had even been with all that food removal.

The improvements Cindy saw were rather typical from the reports. This vast array of progress was a bit surprising at first. Something was going on. Something reaching much farther than just being able to add a few foods back into the diet. Some parents were watching their children just burst to life. For others, it was slower yet steady progress.

But, like storm clouds lingering in the distance, there was that ever-echoing question hovering over it all: Would it last? And was Cindy's son alone as he steadily held on to his gains? Those uneasy about this new enzyme therapy, or wishing it would go away, were waiting quietly, and not-so-quietly, for the fall.

Nutritional Supplements – The science behind it

What Cindy was not alone in was giving a truckload of nutritional supplements! One of the first things I noticed when I started looking into treating these conditions was the bucketfuls of supplements that parents were giving their kids. What was all this stuff for anyway? Giving calcium was easy to understand because if you take out all dairy, you need an adequate source of calcium. I realize that foods such as spinach have a lot of calcium, but I could not see my boys realistically eating five cups a day in order to meet their nutritional requirements. I sure wasn't going to eat it – and is that five cups of cooked mushy spinach or the fluffy green leaves? Taking out gluten meant you needed better sources of magnesium and molybdenum. Many B vitamins are lost there too. Figuring out and rebalancing nutrients is a serious concern with restrictive diets.

From reading the literature on autism, I discovered that many of the biological problems were either thought to be caused by nutritional deficiencies or caused deficiencies themselves. So, wanting to be a good parent, I slowly started adding in all these things like everyone else. The trouble with trying to read about supplements, especially from places selling them, is that they *all* sound vital and essential. Like you must have each and every one to be able to function at all. I was giving handfuls of stuff every couple of hours. This definitely did not feel right. It was also quickly becoming very expensive. Most parents gave anywhere from 15 to 25 different supplements or more. At a good $10 to $25 a bottle per month per child, you can see why parents joyously hoped enzymes would work. Digestive enzymes are a very important alternative. Some families just do not have the circumstances or resources available for other measures.

Everything imaginable seemed to help at least some people, but nothing helped the majority. The inconsistency only made it that much harder for parents to choose the right supplements to give. What a mess! Maybe this is because some children had gastrointestinal, malabsorption, and deficiency problems so great they would improve on anything.

One autism organization compiled results from informal parent surveys conducted since 1967, submitting more than 18,500 ratings on the measures the families tried. It includes many prescription medications and different restrictive diets as well as supplements (ARI 2000). It gives the results in three categories: children got worse, got better, or there was no change. For every option, there were children in each category: better, worse, no change (except there were no 'worse' ratings for melatonin). For most therapies, 'no change' looked to be around one-third to one half of those trying any alternative. So, no matter what you tried, you have some chance of seeing improvement or regression, and around a good 30 percent or more chance of seeing no change at all.

Having a fundamental digestive or malabsorption problem (from any cause and for any person in the general population) can lead to across-the-board nutrient deficiencies. An interruption in nutrient uptake leads to disruptions in the many biological processes requiring those compounds. Harbige (1996) reviewed how nutrition can have profound effects on immune functions, resistance to infection, and autoimmunity. However, although nutrients can enhance the immune system, too much of many nutrients can depress immune function.

Wecker et al (1985) showed that an imbalance in trace elements could disrupt neurotransmitter function and produce significant changes in behavior, many resembling symptoms of autism. When looking at hair analysis for nutrient content, they found clear deficiencies of calcium, copper, zinc, and chromium. A number of malfunctioning pathways and processes have been identified in various children and adults along the way as well.

How do you suppose our little guys became deficient in so many nutrients to begin with? Well, yeast and bacteria could be using those nutrients, gobbling them up as they come into the intestines. Leaky gut, inflammation, and immune system reactions can distort the normal flow of nutrients absorbed. An overburdened liver may not keep up with the demand for the raw and processed materials needed by the rest of the body. Any lack of nutrients can create problems in the hormonal system, nervous system, brain function and sleep patterns. We can consume all the nutrients needed, but if they never reach the right locations, they will not do any good. Whenever gut injury is present, you risk malabsorption. And when you have malabsorption, giving enzymes is wise, at least until the problem is corrected and the gut is fully healed (Friess *et al* 1993).

Essential fatty acids

Quite of a bit of research points to the benefits of essential fatty acids in the diet, especially the omega-3 fatty acids. Although all the essential fatty acids are...well...essential, the omega-3s are particularly important because most people eating a typical 'modern' diet with lots of processed foods do not get enough of these, whereas they usually do consume enough of the other fatty acids.

Fatty acids are critical to proper brain growth and function. The brain is 60 percent fat or more, and accounts for 20 percent of the brain by dry weight. All cells have membranes composed of fat layers. The essential lining (myelin) that coats all nerve cells is over 70 percent fat. Myelin surrounds nerves like insulation surrounds electrical wiring. We need lots of healthy myelin for proper nerve signal communication.

Supplementing omega-3 fatty acids may have significant benefits for neurologically based conditions. Research shows that these fatty acids have effectively reduced hyperactivity in children with attention deficient disorders, brought people out of clinical depression, and also stabilized those with schizophrenia (Edwards *et al* 1998; Horrocks and Yeo 1999). Other research supports that essential fatty acids correct sleep and hyperactivity in many types of conditions including ADHD, autism, and bipolar depression. Much of the research in this area is outlined in the book *The Omega-3 Connection*.

Anecdotal evidence by parents also supports that the addition of omega-3 fatty acids given to their children with AD(H)D or autism has improved some symptoms. One study looked at the fatty acid levels in the blood cell membranes (plasma phospholipids) in children with autism. On average, the total levels of omega-3s in children with autism were about 20 percent lower and the levels of the particular fatty acid DHA were 23 percent lower than in the controls (Richardson and Ross 2000; Vancassel *et al* 2001). Horrocks and Yeo (1999) found that the inclusion of plentiful DHA essential fatty acids in the diet directly improves learning ability and is deficient in people with a number of neurological conditions. Richardson and

Ross (2000) found the fatty acid deficiency to be a link which connects many neurological conditions together – AD(H)D, dyslexia, dyspraxia, language delay, and the autism spectrum.

Essential fatty acids influence how our brains use the nutrients, and affect the function of the neurotransmitters such as serotonin. They are also necessary to maintain integrity of the gut wall. So we need to consume, digest, and absorb these 'good' fats so our brains, guts, and everything else will function properly. 'Bad' fats are fats that are not the right shape or configuration to work optimally in our bodies. The good news is that if bad fats enter the system and cause damage, we can reverse this damage over time. The good fats include omega-3s and omega-6s (including sources such as omega-3 containing eggs), and non-hydrogenated oils. Getting a good quality brand is important because oils can spoil easily and lose their nutritional value if processed incorrectly. Coconut oil has two other outstanding benefits: anti-viral properties and higher heat tolerance.

The key to fats is consuming the right type in the right balance. Do research the subject before you start wolfing down large quantities of only one type. Balance is essential…like the fatty acids.

So where do enzymes fit in? It could be that some people are already getting more omega-3 and other beneficial fatty acids from their diet than they realize. However, if the fatty acids are not digested and absorbed properly, they will pass right through your body. With an injured gut, intestinal lipase for fat breakdown may be limited. If you decide to supplement with essential fatty acids, you may need to include enzymes to facilitate making them available for use. Other enzymes enable the rest of the body to function well enough to absorb, transport, and use those fatty acids.

Secretin

When the stomach acid enters the small intestine, the hormone secretin is released by the small intestines (duodenum). One of the primary tasks of secretin is to signal the pancreas to release pancreatic juices into the intestines. This fluid contains digestive enzymes and bicarbonate. The bicarbonate raises the pH of the small intestine. The higher pH helps the digestive enzymes to work more effectively.

Remember that the pancreas tends to release just enough bicarbonate to neutralize the amount of acid coming from the stomach. Various causes may decrease gastric acid production, including natural aging. Whenever lower quantities of stomach acid enter the intestines, there is a decrease in the release of the hormone secretin. The reduced amount of secretin may then throw off any other function it is tied to, particularly digestion. Secretin therapy may be a round-about way of getting digestive enzymes into the intestines and functioning well.

If something is inhibiting or blocking the pathway, the stomach could be signaling the body for more gastric acids to start proper digestion but it never arrives. The lower stomach acid may not be activating the pepsin in the stomach and food is not well digested. This allows more insufficiently broken-down food to enter the intestines. It may also lower the quantity of enzymes sent to the intestines. On the other hand, if gastric acid production is just fine but something else is hampering the release of bicarbonate, then the very acidic contents of the stomach go into the intestines where they remain acidic. Then acid is in the intestines, animal (human) digestive enzymes are inhibited if not outright deactivated, food is not broken down, the intestinal lining gets 'burned,' not a pretty situation.

Initially, secretin was credited for helping three children with gastro-intestinal problems as well as giving improvement in behavior (Horvath *et al* 1998). However, many, many following studies with much larger groups of children showed no significant improvements with secretin, or the placebo worked as well (e.g. Sponheim *et al* 2000; Carey *et al* 2000; Heisler 2002; Owley *et al* 2001 with 56 children; Roberts *et al* with 65; Molloy *et al* with 42).

It might be the secretin was producing positive results in some by aiding digestion. This may explain some of the discrepancy in results. Not all children would have the same degree of gut injury or digestive response. It could be the added secretin did not stimulate enough bicarbonate production in some, or even if there was more bicarbonate, the timing may not be right to improve digestion or intestinal health. The research of Horvath in 1999 supports this possibility. Of 36 children with autism, 75 percent showed an increase in the amount of pancreatico-biliary fluid output after intravenous secretin administration. The majority of children, although not all, did show greater pancreatic fluid production with secretin. So although supplementing with secretin stimulates pancreatic fluid production in most people, that is only the first step. How effective any increase in fluid is would be dependent on other factors, such as the natural digestive enzyme content, bicarbonate content, if it happens to hit the digestive cycle at the just the right time, if inflammation or leaky gut is present disrupting digestion and nutrient absorption farther along the line, or a multitude of other things. Another view feels any progress seen with secretin may be due to the carrier ingredients in some secretin products, and not the secretin itself.

Giving enzymes directly could be doing what this aspect of secretin is accomplishing only doing it in a more efficient, direct, reliable and far less expensive manner. Digestive enzymes might also clear out any insufficiently digested foods having an adverse affect on stomach acid. Oral enzymes may promote gut healing, an increase in nutrient availability, and natural enzyme production.

If the problem is too little stomach acid, medical professionals may recommend taking supplements such as hydrochloric acid.

If the situation is too much acidity in the stomach, taking 1/8 to 1/4 teaspoon of baking soda, the plain baking soda most people have in their kitchen cabinets, *after* eating may help a great deal. Bicarbonate is the foundation of most antacids and several ulcer formulas. I fill empty capsules with baking soda and keep them on hand. Putting the baking soda in a little liquid or food will also work. This helps with reflux, sour stomach, nausea, and other uncomfortable effects from eating. Some say this in itself can help digestion and overall health tremendously (and the price is right!).

Serotonin

Tryptophan is an important amino acid. It is converted to other forms before finally becoming the neurotransmitter known as serotonin. In a 1996 study, McDougle *et al* looked at the effect of tryptophan depletion in adults with autism conditions. Tryptophan depletion led to a significant increase in behaviors such as whirling, flapping, pacing, banging and hitting self, rocking, and toe walking. In addition, patients were significantly less calm, less happy, and more anxious. Correcting a serotonin deficiency has improved similar behaviors in children with attention deficit hyperactivity disorder as well (Bhagavan, Coleman, and Coursin 1975).

Serotonin works in the brain and gut, helping to regulate sleep and mood. Serotonin amounts are very irregular in many people with autism conditions. Research finds some have elevated serotonin levels, some have levels too low, and some are in the normal range. This explains why some people do quite well on the SSRIs and others become much worse. SSRIs are prescription medication selective serotonin reuptake inhibitors such as Zoloft, Prozac, or Paxil. The SSRIs increase the amount of serotonin activity by reusing the serotonin present longer, not by adding 'new' serotonin to the system. If you start out as too low in serotonin, the SSRIs tend to help, but if you start out with a naturally elevated level, the SSRIs may only make it higher. At the moment, there is no direct or conclusive test to see if the amount of serotonin in your brain is high, low, or normal to begin with. SSRI medications are some of the most effective in benefiting autism conditions.

Oh, it turns out that the vast majority of serotonin production is in the digestive tract. Any disruption there will naturally impact the amount of serotonin produced. Isn't it interesting how very often we start looking at something in the brain or behavior and end up back in the gut? There is a shadowy trend in the general population that many more people are eating a diet deficient in enzymes, and many more are being treated for depression and benefiting from the SSRI medications. Part of treating depression is adjusting the diet to a more 'healthy' way of eating.

Eating more foods favoring tryptophan uptake can affect moods too in the short-run or may help mild depression, but probably will not be the total solution for more severe cases. Tryptophan favoring foods help promote calmness and focused attention. These foods include turkey, dairy, eggs, and whole grains. Meats and other protein sources have high amounts of many amino acids, but the other amino acids outcompete the tryptophan so the net amount of serotonin finally produced is lower. Notice that if you eliminate dairy and whole grains, you lose significant sources of tryptophan which may throw off serotonin levels. Craving dairy and breads may be due to the body trying to increase serotonin levels.

Melatonin

Melatonin is a hormone needed for proper sleep. Zinc is necessary for the production of both serotonin and melatonin, and the prevalence of zinc deficiency in neurological conditions may contribute to this. I find it interesting that my family experienced excellent results from using the prescription medication amitryptiline. It is used often at low doses for sleep disorders, migraine, and general pain. Amitryptiline helps induce the deep sleep that many people cannot achieve on their own. Some parents have found that supplementing with melatonin directly does wonders for their children most probably for the same reasons that my family sees such benefits from amitryptiline: the improved sleep leads to better pain regulation, which reduces sensory sensitivities and results in improved behavior. It all rolls together. However, as with all things, melatonin may produce adverse reactions in some, and there is some concern over the long-term effects of giving a hormone directly and often, especially because the effects on children are not known. There are other medications or avenues to help with sleep as well.

By whatever method, ensuring a good night's sleep should be a priority (and a regular sleep/wake routine). You may be pleasantly surprised to see many other problems simply evaporate, or at least be greatly reduced.

Playing with dominos

Any gastrointestinal dysfunction is going to affect nutrient uptake. And anything affecting nutrient uptake is going to affect all the processes in the body relying on those raw materials. One idea is that you can treat these deficiencies by supplementing each and every specific nutrient separately. You can certainly go out and try giving 20 or more things each day, as well as eliminate 39 different main foods. Supplementing separately may only fix one part of the total

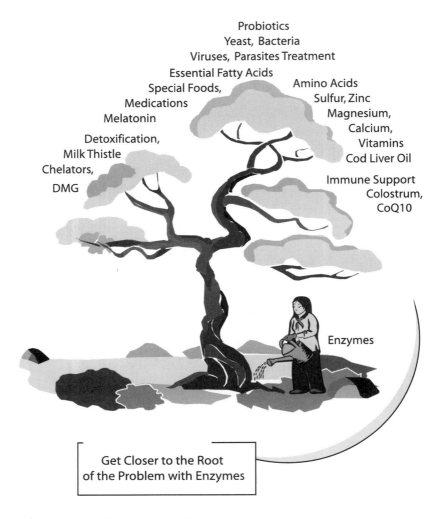

Probiotics
Yeast, Bacteria
Viruses, Parasites Treatment
Essential Fatty Acids
Special Foods,
Medications
Melatonin
Detoxification,
Milk Thistle
Chelators,
DMG

Amino Acids
Sulfur, Zinc
Magnesium,
Calcium,
Vitamins
Cod Liver Oil

Immune Support
Colostrum,
CoQ10

Enzymes

Get Closer to the Root
of the Problem with Enzymes

problem at a time. Perhaps like trying to randomly restack one domino at a time in an entire line of fallen ones.

Another idea is that you can start with enzymes and provide most of these nutrients through a well-balanced diet...with food, and maybe a few select supplements. You deal with the problem at the root level. When we see a wilting plant and want to add water or fertilizer, it is best to apply these at the root so the nourishment is pulled up through the plant restoring health to the many parts. It is far less efficient or effective to water each leaf.

In giving digestive enzymes, you accomplish two major things. First, the person is able to consume the vitamins, minerals and other healthful nutrients from their food. We know that nutrients perform best when consumed from natural foods (Bengmark 1998; Rogers, Zeisel, and Groopman 1993). The delicate and subtle interactions among all the components in a food are not well understood and science has yet to copy them. Isolating out individual components may not do the same thing as in whole foods.

Even if higher doses of supplements are given, they may not be absorbed well in a compromised gastrointestinal system. Which brings us to the second point. Giving enzymes may proactively assist in restoring gut intestinal health at the same time. The domino effect works both ways. One dysfunction slowly drives the system toward ill health, but we can turn the tide on nutritional deficiencies and the dominos fall successively toward health. This further explains the jump in improvement and progress we were seeing in our children and ourselves with enzymes. Especially across so many areas and for so many different symptoms. We were probably correcting many of the dysfunctioning pathways at the same time.

Some found other supplements started to be a source of negative reactions as well. This may be a good sign the enzymes are doing their job of correcting some nutritional deficiencies and malabsorption. However, some metabolic dysfunctions may be fundamentally genetic in nature and need special treatment. Working with an appropriate health practitioner may help to sort these issues out.

This may have something to do with a trend I noticed with my sons. When we started enzymes, I needed to ensure that the boys took an entire capsule of Peptizyde with every instance of casein consumption. If I forgot, the migraines and whining and fussing would start about three hours afterwards. However, during the fifth month, Matthew attended a birthday party with the usual ice cream and cake. We forgot to bring the enzymes and decided he could go ahead and have the treats along with everyone else, and we would just endure the consequences for a few days. Well, the consequences never came. No adverse reaction at all. This happened accidentally a couple more times over the next few weeks. Healing an injured gut? Correcting nutrient deficiencies or dysfunctional pathways? Whatever was going on, it became apparent that from the fifth month on, both Matthew and Jordan could have a casein-containing meal or snack without enzymes about once or twice a week without problem.

I was so impressed by this, I thought I would try it myself. One ice cream sandwich with no enzymes. By the next morning, I had a roaring migraine! I am not sure what makes me so special, but I still needed to take a Peptizyde when consuming dairy – maybe it has something to do with old dogs and new tricks!

By the ninth month, Matthew and Jordan could have several meals or snacks each week without enzymes. This was very encouraging to me because previously I just accepted that they would always need enzymes. Now I look forward to the possibility that they will reach a point where enzymes are no longer required at all. Because of all the benefits of taking enzymes regularly, I compare it to taking a multivitamin…something you just take regularly for general overall health and maintenance. However, it is very nice to know it will not cause a health or emotional crisis if we miss enzymes with a meal or snack. It is wonderful to not be stressed out whenever food pops up.

Others are also reporting they do not need nearly as many enzymes after awhile either, or can better tolerate infractions from their restrictive diets occasionally without enzymes. Usually with unplanned infractions, the intensity and duration of any adverse effects were considerably less after starting enzymes.

Among those in the enzyme group, most parents say they are able to return to giving most foods with enzymes. This is very helpful because it is often difficult getting nutritious things into children. Some of the most frequent questions in our group concerned how to get the enzymes into a bunch of children who are by nature extremely picky and have eating problems. And when do you give enzymes? Did it matter how much? And what happens if…?

As you can tell, as the months rolled by, my boys were doing wonderfully, and so was I. However, some were finding enzyme therapy to be a much more turbulent sea to navigate.

CHAPTER 11

Guidelines for Giving Enzymes
– Getting Best Results

Timing

There were posts from one mom in particular named Lynn. Lynn would say one day the enzymes appeared to work great with her son, but the next time there were disappointing results with her boy acting as if he had not had any enzymes at all. This bugged me. It bugged me because enzymes are one of the few supplements where their exact mechanism is known. I could understand getting either positive results or negative results, but not inconsistent ones. There isn't that much to it. Match the enzyme to the food, take the enzyme when you eat, and it either works on the food or it doesn't. This may be a rather simplistic way of putting it but even if the results were marginal, wouldn't they be consistently marginal? Enzymes see each meal or snack as an independent event. Lynn would have glowing results one time, but depressing reports another. Then a few other parents mentioned inconsistent reports as well. Argh! I would lie awake at night wondering what would cause inconsistent results, mulling over the possibilities. I like consistency.

Then something else happened. My two boys and I happened to be at a local amusement park where a picnic dinner was served. About 20 minutes after eating dinner, my younger son spit up. He would gag up from time to time anyway, so this was not unusual. However, this time while cleaning up, I noticed the intact enzyme capsule. Needless to say, I was a bit alarmed because I was under the impression the capsules dissolved within minutes after swallowing them. This capsule was not even partially dissolved – it was quite intact. Within two weeks, this happened again. An intact capsule gagged up along with the stomach contents. This along with Lynn's frustration got me thinking about when WERE those capsules going to dissolve anyway.

So, I started doing a series of tests in my kitchen dissolving several brands of supplements that came in gelatin capsules, and several brands that came in vegetable capsules (veggie), at different pHs (levels of acidity), and different temperatures. These included the brand Vcaps and common supplements from several popular and reputable companies. After several days of dissolving dozens upon dozens of capsules, the conclusion of my tests was that the veggie capsules did not dissolve as readily as the gelatin capsules at the pH and temperature of the stomach. I even tested this with different foods to mimic stomach contents, over and over. Sitting there watching capsules dissolve gave some much better indication of what was going on. The veggie capsules would pop a leak, and release some of the enzymes, but then the capsule would fold back on itself and reseal. A few minutes later a little more of the capsule would dissolve away releasing a bit more enzymes, but then again the capsule would fold back on itself and reseal. This slowed down the release of the bulk of the enzymes a great deal. At body temperature, the gelatin capsules dissolved away in a few minutes. Poof! They readily disintegrated.

Since enzymes need to be in contact with the food in the stomach to work, I concluded that the veggie capsules might not be dissolving in time to completely release the enzymes for maximum results. I posted all my tests and results with the suggestion that we start waiting 20 to 30 minutes after swallowing enzymes in veggie capsules before eating. After this, Lynn and all the parents getting inconsistent results changed their timing or started opening the veggie capsules and mixing the enzymes with something first. All of these parents achieved consistent, positive results from then on.

A few individuals from another organization argued that their official lab tests showed that the gelatin capsules and veggie capsules dissolved pretty much at the same time. They felt there was no need to wait and felt this finding somehow reflected badly on their products in veggie capsules. (I had thought others might be pleased because this might help get better performance from other enzyme products.) For most supplements, timing isn't that critical. The lab data showed that the veggie capsules dissolved about the same rate, and even a bit faster, than the gelatin capsules. However, the test results did not include the important measure of temperature!

Temperature does not appear to affect the veggie capsules nearly as much as the gelatin capsules. Think of Jello™. Jello™ will last for weeks in the refrigerator but will dissolve very rapidly if placed in the sun or the 98°F (39°C) of body temperature. I noticed from my tests that the gelatin capsules dissolved significantly slower at room temperature and significantly quicker at body temperature. Also, the official lab tests that were given showed that the veggie capsules were not 50 percent dissolved after 15 minutes. Add in that the tests may not have included other 'stomach contents', and you have a good argument for waiting at least 20 to 30 minutes for the veggie capsules to sufficiently dissolve anyway.

What plays out in real life with actual people is sometimes different than what you can derive from lab results. Some people found they get significantly better results by emptying the enzymes out of the capules. The ability of the capsules to dissolve in the gut appears to vary by person. If someone finds that waiting this amount of time with veggie capsules produces better results, then this is what they should do. If you feel not waiting gives better or just as good results, then by all means, go with what works best for you.

By the way, after about six weeks on enzymes with the adjusted timing, my younger son no longer gags up. His reflux ended.

Because gelatin capsules dissolve within two to four minutes under stomach conditions, there is no need to wait before eating. If you are comfortable with gelatin capsules, you can buy a bag of empty ones in a health food store and transfer the enzymes to them if you do not want to wait for the veggie capsules to dissolve. This allows swallowing the capsules at the beginning of the meal or snack. Only a few parents are taking this route, but they say it is effective and convenient. I keep some enzymes transferred to gelatin capsules on hand for when my boys go to a friend's house, in case snacks are served. They keep them in their pockets and just swallow them if something is offered.

If you are opening the capsules and mixing the enzymes with food or drink, just eat or drink this at the beginning of the meal. No waiting is necessary. You can also mix the enzymes with apple sauce, honey, or whatever, and eat those few bites first. If you give enzymes in a drink, have the person drink first before eating, at least one big drink, and then finish drinking the beverage along with the meal.

Most people describing how they use enzymes say they open the capsules, mix the enzymes with food or liquid, and give at the beginning of the meal. Some do this for convenience and some do it because their child does not readily swallow capsules. Giving a good drink (at least a few swallows) of liquid with no enzymes in it after consuming the enzymes will help wash all the enzymes down. Enzymes settling on the skin, like around the mouth area, may 'sting' a little or cause a red sensitive area, particularly with the proteases. This is because enzymes work on dead tissue and so they break down the thin dead layer of cells on the mouth surface, which exposes the fresh underlying tissue. This tissue is not harmed but can be sensitive. You might see a red rash around the mouth when the child swallows a liquid containing enzymes and some of it dries on the outside skin. Parents have found that simply wiping the child's mouth after drinking, giving the child a swallow or two of non-enzyme liquid, or having the child use a straw takes care of this.

Dosing

For general digestion, enzymes are dosed by the quantity and type of food, not by age or weight. You may need to experiment a little and see how much of a particular enzyme product is needed for specific foods for a particular individual. Check the serving size or dose given on the label and use that as a guide. Then watch the reaction and adjust the dose accordingly. Since different products can have vastly different activities, you may need to take four to six capsules of one product to equal the effectiveness of one capsule of another product (some calculations are in the Guide to Comparing and Buying

Enzymes, Appendix A). The goal is to give an amount of enzymes that can break down all the food eaten in the 60 to 90 minutes the food and enzymes are together in the stomach. Although the enzymes will continue to break down foods as the stomach contents move into the intestines, any casein, gluten, carbohydrates, or other foods not adequately broken down in the stomach and passed into the small intestine is a candidate for incomplete breakdown and problems in the rest of the gut.

Here is the technical reason why: Enzymes are 'inactive' until they get wet and are in contact with the type of food they act on (the substrate). Then the enzymes are activated and go to work. They continue to work until they run out of substrate, or something else happens to deactivate them. Enzymes by definition facilitate a reaction and are not used up or changed themselves in the process.

As an additional scrap of information, enzymes produced by the pancreas or stomach do not start digesting the body because they are not activated until later (an example is the hydrochloric acid necessary to activate pepsin). The body has a series of checks and balances to activate and deactivate enzymes so they are not acting on healthy and necessary tissue.

Since pancreatic enzymes are destroyed by the stomach acid, pancreatic enzyme supplements are usually encapsulated somehow to protect them so they will survive the stomach environment and hopefully be released in the small intestine. There they come into contact with the food and start working. The strategy of using plant enzymes is different than pancreatic ones. With pancreatic enzymes, you do not take advantage of digestion beforehand in the stomach.

If we had hours of time for digestion, we could take a little bit of enzymes and they would continue to break down the food all day long. However, digestion in the stomach happens for only a limited time. You need to give an amount of enzymes that can break down all of the food in the 60 to 90 minutes the food and enzymes are together in the stomach. Remember the one guy with the master key trying to unlock all the doors in building? So, for best success, take enzymes at the beginning of a meal or snack. This maximizes the amount of time for food breakdown before absorption. The exact quantity of enzymes needed will depend on the food type, the product, and the individual. This can also change over time. As a damaged or leaky gut heals, you may need fewer enzymes than in the beginning. Including the possible increasing contribution of the enzymes in the mucosal lining becoming more functional and available as the villi heal and grow anew.

A conservative way to start on enzymes, minimizing possible adjustment effects, would be to begin with a general enzyme formulation that is low in proteases. You may want to start low and gradually increase the dosage. Start with about a half capsule. If this causes excessive hyperness or undesirable reactions initially, drop the dose to less than a half capsule and build up slowly, going even as low as one-sixteenth of a capsule, or just a few sprinkles. If all goes well after a few days, increase the dose little by little until you reach one capsule. You can increase up to more capsules per meal, depending on the product, or if you feel giving more might be beneficial. Then add the strong protease product, increasing the dose slowly in the same way. Gradual dosing may be helpful in reducing any hyperactivity and other adjustment effects, especially with proteases and especially if the gut is severely injured.

Check the 'serving size' and activity levels on each product to get a general idea of what to take. Be sure to read the entire label carefully. Most people do fine with using just one capsule, maybe two, of Peptizyde per meal or snack. More than three capsules will most likely not make a difference in digestion. If you need more than three capsules of an enzyme product to eat a specific food, you need to weigh the benefits of that ability versus the cost of the enzymes. If this is your situation, you may want to just select some meals for special occasions with enzymes, and eat restrictive diet meals the other times.

For getting benefits of enzymes other than digestion, there is a general 'therapeutic' maintenance dose of around nine capsules or doses per day (some products say 2 or more capsules equals one dose). Taking proteases between meals may be effective for health benefits besides direct food and supplement breakdown. For specific cases of treating an autoimmune condition, you would need to research each particular condition and find out the dosing and enzyme mix needed. One product I saw said to take 45 tablets of that product per day to help fibromyalgia. Certain mixtures convey benefits where giving the enzymes separately do not. It can take between a few days and a few weeks or months to start feeling better. For sports injuries or similar situations, taking more proteases per day for several days seems effective, in general.

For healing a damaged gut, consider that the gastrointestinal system is a dynamic place. It starts repairing any damage immediately. The estimated time for gut healing is typically between three and six months. For a more severely injured case, healing a leaky gut can take up to 12 to 18 months.

Grazing

Grazing is when a person does not sit down and eat three separate meals a day, but rather eats a little continuously or snacks all through the day. So when do you take the enzymes? Here are some strategies that can work with grazing. One is to mix the enzymes in a cool drink and have the person take a good drink or two each time before eating. The enzymes will stay active in the drink for up to four hours if it is kept cool. Enzyme activity is directly related to temperature. The lower the temperature the slower they work. You can keep it in the refrigerator or add ice to it (or both).

Another method is to sprinkle some enzymes on a bite of food right before the person eats. The capsules usually open quite easily by pulling them apart. You can sprinkle some enzymes out and then snap the capsule back together until you need the enzymes again. Another idea is to make the candy or frosting wafers, and take one each time before eating or snacking (see Mixing Suggestions below, page 165).

I found I felt much better during the day by putting enzymes (usually the Peptizyde or proteases) in a drink and sipping on it all day. I used an assortment of different drinks. The idea of an 'enzyme slurpee' became popular especially with people who graze. When I slacked off and only took around three or four capsules per day with meals, I really noticed the difference, feeling less energetic and more pressure in my head.

Frequency of enzyme use

Many parents who used enzymes occasionally and then switched to using them regularly with all foods are seeing significant improvement when they use enzymes all the time with all foods, especially in the beginning. This tends to bring an increase in the Happy Child Effect. The child becomes noticeably more pleasant, easy going, cooperative, and helpful. There seems to be a synergistic effect over just giving some enzymes some of the time or occasionally.

Even when good improvement is seen when giving one product or using enzymes occasionally, some people see far more improvement when all food types are addressed with enzymes all the time (at least in the first month or two). This is probably due to the other healing properties and benefits of digestive enzymes besides just the immediate need of breaking down particular foods. Possibilities include:

- gut healing and colon cleansing with enzymes
- immune system support
- blood cleansing or purifying
- improved sleep and mental capacity
- weight regulation (underweight people tend to gain and overweight people tend to lose)
- balancing gut flora; cleaning out yeast, parasites, virus, and bacteria, and their associated toxins

What if you are on a restrictive or other food-free diet and wonder how often you can have an infraction? This is up to personal choice and perhaps individual sensitivity to particular foods or chemicals. As far as the enzymes are concerned, they work in a systematic way on specific foods. Enzymes see each meal or snack as an independent event. Having a snack with enzymes on a Tuesday does not affect how enzymes will act on a meal on Friday. An infraction on Wednesday morning will not effect having another infraction Wednesday afternoon, from the enzymes point of view. Some people found they leave a restrictive diet almost by accident. Infractions came closer and closer, ever more frequently until one day, they realized they were no longer restricting foods at all. Gut healing? Immune system support? More enzymes in circulation may lead to reduced reactions because they may digest any circulating peptides, complex sugars, or other particles in question; or otherwise assist the body in processing these out appropriately.

Mixing suggestions for enzyme supplements

The following suggestions are compiled from ideas offered by parents and other individuals on methods for giving enzymes. Many of these ideas work equally well for other supplements. Parents and/or pharmacists report using the following liquids for mixing enzymes:

- water, fruit juices, milk or milk substitutes
- soda pop, spritzers
- sports drink, lemonade, limeade, powdered drink mixes
- syrup: chocolate, strawberry, cherry, other flavors
- a masking or base solution may be available from your
  local pharmacy or compounding store, as well as other sources

Foods parents have used to hide supplements:

- applesauce, pear sauce
- chocolate syrup or something similar
- honey
- frosting, icing
- frozen fruit concentrate
- ketchup
- oatmeal, mashed potatoes
- peanut butter, other nut butters
- puddings
- sorbet, ice cream, frozen yogurt
- mixed with butter, jelly, jam and served on a cracker,
  bite of muffin, bread, waffle, biscuit

If your child prefers to take the capsule unopened, you may want them to swallow it about 30 minutes before the meal if the capsule is vegetable-based. This ensures enough time for the capsule to break down and the enzymes to make contact with the food. Digestive enzymes need to be in contact with the food to work. This is not an issue with gelatin capsules. If taking enzymes with food, it is also advisable to drink some water or other liquid to wash the contents down to the stomach, and wipe your mouth, if necessary.

There is a big benefit in teaching your child to swallow capsules, if possible. If he or she is over three years old and there is not a potential choking problem, it may just make life easier for everyone. I taught my younger son by putting the capsule or pill on the back of his tongue, then had him take a drink of liquid, then look at the ceiling as he swallowed the drink. Gravity takes over. Now he downs four capsules at a time. Another mom started with very tiny things to swallow, gradually increasing the size until a capsule could be downed.

Does chewing the capsules invalidate the effect of enzymes? I do not know how others feel about this, but I don't think many enzymes taste good. Some have no taste and others are extremely strong. If you take enzymes out of the capsule, make sure you wash them down with a couple of swallows of liquid. Enzymes sitting in the mouth (and this may include chewing) may start to irritate the surface tissue. The 'irritation' is not harmful, just uncomfortable – like when you get a sore or raw spot in your mouth.

Can enzymes be mixed with other supplements? Since most supplements are not substrates for enzymes, the enzymes usually do not interact with them. You should be able to mix most supplements and medicines with enzymes. Check with your doctor about specific medicines, especially time-released ones. Enzymes may affect the properties of time-released medications increasing the rate at which they are broken-down, and thus, released.

There is a question about giving enzymes with probiotics, especially the proteases. Some people see better results when proteases and probiotics are given separately, some do not. So check with others using specific brands, try it both ways and see what works best for you. Probiotic formulations and encapsulation methods can vary widely, influencing results. The safe bet is to give enzymes at the beginning of a meal and probiotics at the end of the meal or at a different time altogether.

Mixing supplements in a liquid for later or goind out

Cindy suggested her method of using pear juice in the four-ounce plastic bottles from the Gerber baby food company. They are perfect to keep in your purse or in the glove box of the car. At a restaurant, she says she usually dumps out or drinks half the bottle, then empties the enzymes into the rest. Then, she puts the lid on and shakes it up. It is a great way to mix the enzymes. She reuses the empty bottles.

Sometimes she pre-mixes the enzymes, then puts them in a lunchbox with a freezer pack and takes them with her if she does not want to deal with pulling out capsules. That is how she handles sending the enzymes in her son's lunch, too. You can buy pear juice or other type of juices in a larger size also and just fill the little saved bottles as needed. Larger sizes are available at grocery or health food stores. Small plastic drink boxes can be bought separately also.

If you pre-mix the enzymes in liquid, they should remain active for about four hours, if kept cool. Enzymes are 'activated' when they become wet – so that is why you do not see bottles of liquid enzymes or want them to be in a liquid days in advance. This is also why enzymes need to be stored in a cabinet and not a refrigerator (moisture may collect with the opening and closing of the door).

If you were going out, you could mix up a batch ahead of time, keep it in the refrigerator or freezer until you leave, and then take out with you. Putting this in a drink bottle to take for school lunch, a restaurant, or shopping trip works well. Also, you can use a bottle with a dropper to take out away from home, and dispense mixed enzymes in this way. Oils do not have this same effect of activating enzymes, so mixing enzymes with peanut butter, or the frosting or chocolate chip ideas, work fine.

Enzyme chocolate wafers – Mixing enzymes in a solid for later

First, melt a few chocolate chips (or something similar your child can tolerate) in a small bowl in the microwave; be careful not to scald. Make sure you do not microwave the enzymes themselves because high heat destroys them. The chocolate should not be too hot before you add the enzymes. A good test is to see if it is cool enough to put in your mouth. Anything cool enough to eat is fine for the enzymes. When ready, dump in the amount of enzymes you would like for a 'dose' and any other supplements you want. Mix, and then transfer to waxed paper and freeze. The mixture will freeze in about two minutes. These will last up to two weeks. The person can eat one at the beginning of a meal or snack. They are fine to eat frozen, or sit out to 'thaw.' One boy called this his chocolate cookie. No more battles and much less waste. These are good for school, vacations, and traveling. One mom said she did this instead of packing all her supplements in a suitcase for trips. She mixes up everything in chocolates. These have been very popular with both parents and kids.

Advantages of using a straw for drinking liquids
Using a straw may help some people tolerate the taste of the enzymes better, if this is a concern. You do not taste things as strongly by keeping the straw toward the back of the mouth. Some enzymes have a very mild taste or no taste at all, whereas others have very strong tastes and smells.

Using a straw will help keep the enzymes off the skin. Otherwise, be sure to have your child wipe his mouth after drinking the enzymes from a cup. If you notice a slight reddening or rash around the mouth, this is usually the cause.

Other tactics for getting enzymes and other things in kids
1. Mix the enzymes with honey and spread it on a small corner of a cracker, pancake, sandwich, or whatever, then have the child eat that piece first. Can try this with butter, nut butters, jelly, jams, etc.

2. Give the child a choice: either put the whole capsule on the back of the person's tongue a few minutes before eating and wash down, or eat it mixed in the first few bites of food. Offering a choice gives the person some control over the situation.

3. As for food, you can put the capsule on a spoon with a little pudding, chocolate syrup, or something similar and let him 'eat' it that way. Do not try to hide the capsule, just say he needs to take it that way, but he can choose the food. This gives him a choice. Taking the capsule is not a choice, choosing what to take it with is his choice. Since this may become a routine, you might just want

to have a 'talk' with the person and say, 'Look, you need to take this. It keeps you from getting sick. You choose what you want to take with this.'

4. Putting a little butter, jelly, or similar stuff around the capsule makes it 'taste' a bit better, and maybe it will slide down easier. One boy liked to coat the pill with peanut butter. Even dipping the capsule in plain water may help it slide down better.

5. Will your child refuse to eat if you tell him no food without first taking the enzymes? If he does refuse, will letting him go hungry for a bit until he gets hungry enough and take it work? You may have to wait him out awhile, but most parents have not had a problem after one, maybe two times of this. However, if you feel that this is not benefiting your child, do abandon it. Some children will go to the point past reasonable, and another tactic needs to be taken. The feeling 'he will not starve' does not always hold true. Some children may not have the inherent neurological feedback signals working properly that make this tactic a success in typical situations. Perhaps withholding computer time or something else will work.

6. Do rewards work?! Cindy's son has a reward system using colored links that connect to make a chain. She starts it on the refridgerator door and when the chain touches the floor, he gets a reward.

7. If you need to get the enzymes in the person fast, open a jar of pears or other fruit, mix the enzymes in and eat.

8. If something does not taste good, have your child hold his nose while he swallows it, either from a spoon or straw, then while still holding his nose take a drink of straight water or juice to wash it down (as a chaser). You can breathe through your mouth while you do this but keep your nose closed. A lot of taste is in the smell. If you use a straw, the stuff with the smell is less likely to get up your nose where it will linger (I do not like fish because it is a little oily and it gets on the roof of my mouth where the passageways are and I smell fish for awhile afterwards).

9. Syringe: You can buy several kinds of infant oral syringes (pharmacy or grocery store), or use one you already have. They may be slightly different sizes. Hold the tube with the left hand with your index finger covering the hole. Pour in whatever liquid you want to prepare and then add in the enzymes and other items. Use a piece of dry spaghetti and mix the solution up. Then get the pusher part. You can push it all the way down by gently moving it at an angle, wiggling to let bits of air out. At this point, you can take your finger off the hole and none of the solution will come out. It's airtight. At first you may need to hold your child still to be able to shoot it all in his mouth, until he learns to just open his mouth. Placing the solution more toward the back of the mouth will lessen any taste. Washing it down with a drink right away also helps. Some kids think this is really fun.

10. Hiding things in food may cause a child to not trust you (not good), and not trust food (also not good), and then downright refuse to eat for fear that something may be slipped into the food. What has worked the very best for some parents is this: When you give a child something yucky, just tell him it is going to taste yucky! Say yucky several times while he's taking it. If it is something he likes or does not mind, you can say yummy! It's so subtle he may like the fact that he can tell the difference. No matter what, he needs to know he needs to takes it. Letting him choose when to open his mouth helps!

11. One parent wrote: My three-year-old cannot swallow capsules and can sniff out anything in his drink. After trying a lot of different methods, I took a gum drop and scraped all the sugar off the outside (it then looks like a gummy mountain – you can leave the sugar on if you like). Then I cut a sliver off the wide end and then gutted/scooped out all but a thin layer of the outside wall. Then I pack in the enzymes and any supplements. I then place the bottom sliver back on. I set it on his plate and he knows that he eats that first, swallows three gulps of his drink, and then can eat. You can let the child choose the color of gum drop he wants.

12. Try a token system. One mom gave her son a bank he carried around. Every time he took his supplements he was rewarded with either a penny, nickel, dime, etc. At the end of the week, he counted his money and bought a little something. He loved it. He loved the bank, the coins jingling, and getting the reward. You do not have to use money. It can be tokens to play a special game or whatever you think will be motivational. Be very excited about doing this. Let him pick out a bank or the tokens.

13. Another idea is to mix the enzyme with some double chocolate cocoa powder. One mom did this: I get one that comes in a can with a resealable lid. I open up the enzyme capsule over a spoon and let the amount I need fall onto the spoon then gently scoop up some cocoa powder out of the can, and swish the two around with my finger to blend it, and then let my 18-month-old have the whole spoonful — dry. Then follow up with some liquid to drink.

14. You can take chocolate or vanilla dairy milk, rice milk, soy milk, juice, or other beverage and chill it thoroughly. Add in the enzymes and immediately freeze into popsicles. This is a fun treat especially in warmer weather. This may not work as well with enzyme products having strong tastes and smells.

15. Because one mother's child was so used to taking supplements by medicine dropper (since one year old), she used the same technique for the enzymes. She put the enzymes in a small shot glass, added about 10 milliliters (about 2 teaspoons approximately) of orange juice and then pumped on the dropper squeeze/squirt, squeeze/squirt until the enzymes were almost totally dissolved (takes about 30 to 45 seconds). Then she shot them in her son's mouth. He swallows them no problem.

16. There is a chocolate sauce that you keep at room temperature but when you put it on ice cream it sets hard straight away. One mom poured a little in a small cup, threw in the supplements, tipped out the mixture on one of those smooth ice blocks from freezer and zip — hard — slide off — in mouth.

17. One mom empties the dry enzyme powder into a plastic sippy cup or drink container with a lid, and adds some dry lemonade or flavored drink powder. She sends this to school with instructions for the teacher to just add water and shake really well before snacks or lunch, and give to her son. Fresh enzyme drink, no mess or hassle.

18. A mother gave this recipe for frozen chocolate balls: I keep two small containers of homemade carob icing in the fridge. I use two different types of enzymes, so I mix one type in one container and the other type in the other container. I add more carob powder to one container making it darker so I can tell them apart. If you only use one enzyme product, you would only need one container. To each container add a few teaspoons each of: water, melted butter or similar, and carob powder – then mix with enough confectioner's sugar or powdered sugar to make a dough consistency. The dough is dry enough that it does not stick, but not so dry that it crumbles. I add in the amount of enzymes I want. After mixing the enzymes and icing, I form it into a pancake and stick it into the freezer. Then I wait awhile before cutting into 16 sections. I roll each section in my hand into a little ball. So if you wanted each ball to ultimately contain an entire capsule of enzymes, you would add in 16 capsules of enzymes to the icing mixture. If you wanted each ball to contain one half capsule in the end, you would add eight enzyme capsules to the icing mixture. This is not an exact science, so adjust it to suit your needs.

   The dough is easier to work with (less sticky) if it is slightly frozen. Each ball ends up being either one-sixteenth of a teaspoon or one-eighth of a teaspoon in size. I place these out on a sheet of aluminum foil and stick them back in the freezer. If the icing mixture becomes too dry, I wet the knife I am using for mixing and this adds enough water to do the trick. If I get it too wet, I add more powdered sugar. If you think you need to add more liquid, also try adding more butter or oil. My kids know to go to the freezer and take one light and one one dark ball every time they eat. For school lunches, I sandwich the two frozen icing blobs between two cookie or crackers pieces and send to school.

The results are in!

With the realization that capsule dissolving time might make a difference and that having celiac influenced results, the members of the group interested in enzymes spread these guidelines to others. It was a simple message of:

- match the right enzymes to the right foods

- watch your timing, especially with swallowing veggie capsules

- if you have celiac disease, do not consume gluten even with enzymes targeting proteins

- check the information on dealing with adjustment effects for suggestions (especially the hyperactivity)

With those few instructions, the reports of negative reactions dropped in half, and the ones of bewildering inconsistency stopped completely. This increased consistency greatly pleased me. We were getting somewhere. A very fundamental but important aspect was to simply write things down. That way newer people could review previous discussions and findings, and add to it if they liked. I was very pleased to at least target certain subgroups and have guidelines. At least one alternative, enzyme therapy, was a little less chaotic.

As we rounded out discussions on particular topics, the information was summarized and made available as 'help guides.' Making these assessible was important because of our belief that individual parents are the primary caregivers and experts for their children. They are the ones needing information, making decisions, observing behaviors, and trying different measures. One of the first needs was how to read enzyme product labels, so you know what is in it, how much it was costing (or saving!) you, and which foods to use it with. Appendix A is a Guide to Comparing and Buying Enzymes and presents seven steps to navigating the world of enzyme supplements.

On top of all this, Cindy and I got a crash course in managing human dynamics. Keeping things open yet focused became a full-time job. Each person is very different and the whole area of digestion, neurobiology, food intolerances, nutrition, and expressed behaviors is complex. Fortunately, Mike has a billion years experience in dealing with new developments in biology and technology, and how society adapts to change in general. He helped us understand what was going on with 'change management' and how to deal with it. What started out as two Midwest moms who met on the Internet just really elated over their kids improved health became The Great Adventure. A trip we had not planned to take. This was far more than just having a new product on the market. It reflected a change in therapy, mindset, and approaching a neurological condition altogether. No wonder there was such a fluster. The adoption of new technology, methods, or new ideas follows a well-characterized, predictable pattern. Human nature is pretty consistent in the general sense. Even when it is good, change can be unsettling. Change can be hard.

Finally, the day, or rather month, we had been waiting for came. How had the enzymes fared over time? Cindy and I assessed the positive and negative results based on as many sources as we could find. I contacted anyone who had said they started enzymes but never gave a follow-up. We reviewed it all again. I tallied up the count several more times to be sure, and posted the results for all to see.

Chapter 12

The Happy Child Effect

Along with the side-effects and adjustments with enzymes, there were many positive gains and skills being announced. As more and more reports came in over the months, it became very apparent that most people using these enzymes were seeing positive results. The improvements were sometimes very noticeable and sometimes subtle, but growing over time. Most results were seen very quickly – within three weeks or so.

The four-month mark came. No crashing.

Not one. Hundreds of people were using these particular Houston enzymes by now and not one report of the intense regression that was so famous when leaving the casein-free, gluten-free diet (without appropriate specialized enzymes like Peptizyde, that is). I summarized everything found to date from the Enzymes and Autism group and wrote up a summary report. One summary was written for the first four months, and another one later at the seven-month mark. More details from both reports not discussed elsewhere are in Appendix B, Enzymes in Action – Seven Months of Real-life Use.

As I was tallying results, I counted as positive those messages where the person wrote that enzymes were successful, or contained such descriptions as: great, excellent, miracle, thrilled, remarkable, happy, doing well, improving, etc. Others, such as teachers, doctors, relatives, or neighbors verified most of the positive results within the first three to four weeks. From the first four months of those reporting responses, the breakdown was:

Percentages (0–4 months)
100%   151 Total number of individuals tracked
 87%   131 Number of those with positive results
  8%   12 Number of those with negative results
  5%   8 Number of those with inconclusive result

Some of the positive results were nothing short of miraculous. It was very thrilling to see the wide range of gains children were making. There was one particularly interesting improvement that we just started calling 'The Happy Child Effect.' Following is a description of this effect and other improvements commonly seen in no particular order and using the following scale:

- The term 'most' is taken to mean greater than 75 percent of those who commented on a particular topic (meaning very, very common).

- The term 'many' is taken to mean 35 to 75 percent of those commenting on a particular topic (meaning common).

- The term 'some' is taken to mean 10 to 35 percent of those who commented on a particular topic (meaning not as common but occurring regularly enough).

- The term 'a few' is taken to mean less than 10 percent, (meaning rare but occurring maybe a handful of times). Usually 10 individual people or less. Keep in mind that not all individuals commented on all subjects or symptoms.

Increased eye contact – Most children improved in eye contact. Their attention improved, they responded when spoken to, stayed on task longer, and participated in and followed a conversation or activity better. They were said to be 'bright-eyed' and attentive – not having the previous 'blank' look. A few improved to the extent eye contact is in the 'typical' range. Several parents commented their children now insisted on eye contact from adults. One mom said she was working on her computer when her son entered the room, and continued to work as he talked. After a few moments, the boy said, 'Mommy, you are not looking at me. How do I know you are listening?'

Increased language – Most children improved in language in a variety of ways: better speech, more speech, more vocabulary, more expression, better comprehension, clearer speech, more appropriate tone and speed, speaking at a more natural rate, using language and responding appropriately, dropping video talk, etc. This includes the child going from no language to a few words, from a few words to many words, from short phrases to many sentences, and from sentences to conversation. Children participating in extensive conversations and talking about their feelings came up. Even a couple children who did not speak until the age of six years old opened up attempting to communicate. Matthew always had language, but he started being much more conversational. At times before, he would just bark or growl at us instead of answer. Since enzymes, he always expresses what he wants or needs with words and sentences.

Increase in humor – Some parents report their child displayed an increase in humor. This included telling jokes, 'getting' or understanding a humorous situation, and laughing appropriately. An example of appropriate laughing would be when Matthew started spontaneously laughing at Bugs Bunny cartoons. Before, he would watch and blankly analyze it ('Anvils don't really fall from the sky. And if one did fall on your head it would not push you into a hole in the ground. You would have to go to the hospital. Rabbits don't eat carrots with one hand standing up either.') It is nice that he just enjoys cartoons now.

Improved sleep – Most children having previous sleep problems now sleep soundly through the night. This means most parents are now sleeping soundly through the night as well. Sleep difficulties are somewhat common with neurological disturbances, so this is a very nice effect. Some people see a pattern of their child waking up in the night. This is a characteristic symptom of phenol sensitivity. Usually this would end when the parents reduced the amount of highly phenolic foods these children ate. Epsom salts also reduced this (see Sulfur, Epsom salts, and Phenols Chapter 16). No-Fenol enzymes were not available at this time but now may be very heplpful with phenolic foods. Night waking can also be caused by yeast or bacteria.

Increase in foods tolerated – This refers to a physical reaction to foods. After eating previously problematic foods with enzymes there were decreases in or absence of red ears, red bottom, rash, bloating, distended abdomen, stomach ache, reflux, gagging, circles under the eyes, lethargy, hyperness, tantrums, whining, etc. Most parents say the enzymes greatly helped expand their child's menu and made life much easier. Even though some foods continued to produce a negative reaction, it was much milder and shorter-lived than before. Most parents could return to giving whatever foods they wished.

Increase in foods accepted by child's choice – This refers to increasing the number of foods accepted and eaten without a major hassle, and the quantity of different foods a child would eat. An example is Jordan who went from insisting on only five to six foods to about 25 to 30 foods. This was a very common response, although maybe not as dramatic in all cases. Several children increased the amount of food they ate, the total quantity. Others ate in a more timely fashion, seemed to enjoy their food more, and asked for a wider variety than before taking enzymes. One boy asked for salad and we all fell over in shock.

Weight gain – Some children needing to gain weight did so after starting enzymes. Examples: One child had not gained weight in two years and put on five pounds in a few months. Another child put on two pounds in one month after no weight gain in months. This was accompanied by an increase in appetite. There was also a noticeable height increase for a few children who were behind in growth.

Weight loss and reduction in badgering for food – Some children had a problem with a constant need to eat or chew, and nagged for food or drink throughout the day. For those children, the enzymes seemed to calm this so they were in a normal eating routine. Taking enzymes seemed to take the 'edge' off them and reduce the cravings. This was noted especially for dairy, bread products, sugar, and fruits. Matthew and I tended to have 'the chewies', called this because of a driving need to be chewing on something. Matthew would chew through food, clothes, straws, toys, anything. I would crunch hard cereal, raw carrots, or ice, or constantly have a drink to sip on. This dropped off a great deal with enzymes. It is also considered a sensory issue, which may be another avenue to look into if you deal with the chewies at your home. Overall, the common effect of enzymes regulating weight to a more appropriate healthy level was seen, whether the child needed to increase or decrease their weight.

Stools/bowels – One way to tell enzymes are working is by what you do not see or feel anymore: no indigestion, no gas, no bloating, no heartburn, or no undigested food in the stool. Respondents noted all of these improvements. Many parents said their child's stools were improved with enzymes by being firmer, no longer light colored, more regular, and with a decrease in inappropriate and overwhelming smell. The smell is in large part due to undigested food and waste rotting, putrefying, and fermenting in the gut. Good digestion will eliminate most of the foul smell. A foul stench could also be due to bacteria overgrowth.

Skin color and overall appearance – Some parents say their child's skin is less pale, has good color, with decreases in rashes and eczema, circles under the eyes have lessened, face looks more filled out and there is an overall 'healthier' appearance. A few adults taking enzymes have commented their adult skin conditions have improved as well.

Improved transitioning – Most parents say that before taking enzymes, changing activities was a struggle and usually resulted in yelling, whining, or tantrums (or barking). After starting enzymes, these behaviors decreased dramatically. Transitioning was done more cooperatively even with unplanned, unexpected changes in schedule. Schools, therapies and other activities the children participated in commented on this as well (martial arts, sports, art, playgroups, etc.).

Desire for physical activity – A few reported that getting the child outside was previously a major struggle. Now they comment on the child being compliant and even enthusiastic at least half the time. This may be related to an increase in energy, and a decrease in pain or discomfort. It may also be related to a better tolerance for sunlight, temperature, or other sensory based things.

Decrease in general anxiety – Most parents say their children became less anxious or rigid in general. They became less demanding that things go their own way, could transition much more easily, were far more flexible in activities and temperament, less agitated in general, often allowed others to choose first either voluntarily or when told, and were far more cooperative. Children and adults on enzymes were overall much calmer and easy-going. As an additional note, having enzymes to allow eating out occasionally or just to relax about food contamination issues decreased the anxiety of the parents, and thus improved the overall family atmosphere enormously.

Improvement in socialization – Most parents commented that their child took a positive interest in siblings, playing with them, talking, laughing, and asking about them. This included going from ignoring the sibling completely to playing regularly with them for hours and interacting with them throughout the day. This included chatting during chores, playing games, sharing plans to play tricks on the parents, explaining ideas, asking opinions, and other interactions.

Children initiated conversations and talked about feelings and abstract ideas more. Most parents noted a big improvement in handling social situations outside the home as well. The children were more willing to participate in activities with others, more interested in social activities, and there was more discussion of social activities after the fact. Some with children enrolled in school commented on how much better the school day would go, getting very positive feedback from their teachers and therapists. Some with children in therapy such as speech, auditory therapy, sensory integration, or discreet trial training posted on how much more the child got out of the sessions.

One mom was elated that her son is really wanting to play with other children now. He is actually asking if he can play with them, whereas before he really was not the least bit interested. Then she experienced a new situation which some of us find ourselves in. Her son had to take a time-out at preschool because he and two other little boys were horsing around when they were not supposed to be. The parents were thrilled. The view is this is an improvement in several areas: interacting with other children, participating in a group activity, talking to others, and playing pretend.

Increased awareness – Most parents report a significant improvement in all aspects of awareness. This means their child is more aware of the actions of others, more aware of their surroundings, what is going on around them, the feelings of others, and how to behave appropriately without having to be told several times. The child is not nearly as detached, aloof, or 'in their own world.' They initiate play and conversations with others. They acknowledge when others enter or leave a room. A few noticed a better awareness of safety issues and a decrease in impulsive behavior.

Increase in problem solving – Some parents see that problem solving is no longer a weakness. The child is able to reason better and does not give up as readily as before enzymes. The person is more aware of consequences and act appropriately. They participate with siblings or friends in decisions and problem solving. Older children were noted as being more pleasant and patient when doing their homework, trying to learn, and trying to work through assignments, either on their own or with help. Adults feel they are more productive at work.

Improvement with short-term memory – A few parents say that their child now remembers requests and general information better. Parents do not need to repeat their requests as often. The child can remember and act on a multi-step request and repeat the process more often.

Less adherence to routine – Some parents say their child is significantly better able to handle changes in daily routine, such as wearing different clothes, sitting in a different chair at dinner or watching television, or making an unplanned stop while doing errands.

Increase in range of interests – Many parents wrote their children previously had a very narrow range of interests and it was difficult to get them interested in anything else. Their child picked up a new wider range of interests on their own after starting enzymes. They also would more easily take an interest in something new.

Sound tolerance – Some parents mentioned their child had a sound sensitivity that was previously a problem and this was greatly reduced if not altogether eliminated on enzymes. These children could now tolerate a sibling's voice, noisy stores, loud sudden noises, and an array of noises at the same time. Auditory integration or listening therapy are avenues to consider for remaining sound intolerance. After making huge gains with enzymes, one mom started listening therapy and her child improved by another leap. It also demonstrates how enzymes are usually not the only therapy needed, but work quite well in making other therapies more effective. (the eternal catalysts!)

Light and color tolerance – Some parents commented on how their children no longer reacted badly to bright lights, certain colors, different types of light, and sunshine. A few with migraines and headaches said their head pain was significantly reduced in frequency and intensity. (Including me, but I still can't figure out what possible use there is for the color orange.)

Improvement in other sensory issues – Many parents posted messages that their child improved significantly across the board with sensory issues. These included wearing certain textures of clothes, wearing different weights of clothes, playing with items of different textures without having to 'clean off' their hands, accepting different textures and temperatures of food, and far more tolerance of various environments even if it was a new place they were not familiar with. Some children could tolerate very sensory intense places for much longer without getting overwhelmed as before, sometimes for hours. Most of the time there were just far fewer sensory issues to deal with. Looking into sensory integration therapy can be very effective for remaining sensory issues.

Decrease in aggression – A few parents feeling their child was too aggressive before enzymes say this improved with Peptizyde and Zyme Prime. Sometimes, the Happy Child Effect took place and so the parents saw the aggression go away. Other times, the parents say the aggression is still noticeable although not nearly as frequent, not nearly as intense, and it does not last as long. One mother said, 'My son no longer seems to need to compulsively hit. He can stay calm enough to find a better way to express it or handle it.'

Decrease in hyperness – A few parents feeling their child was too hyper or 'wired' before say this improved. It is common for a person to have an increase in hyperness when starting enzymes too. This can happen for a variety of reasons (discussed in Chapter 9, What to Expect When Starting Enzymes).

Increased affection – One of the most blessed and welcome improvements! Many parents say their children became increasingly affectionate on their own, spontaneously giving hugs and kisses, saying 'I love you,' holding hands, not jerking away from casual touching, and showing genuine concern for others' feelings. These are some of the most priceless benefits. My older son Matthew never wanted to cuddle before and could not stand kisses. Now I regularly get spontaneous hugs and kisses throughout the day. Yeah!

Decrease in self-stimming overall – Many parents report a significant decrease in a range of stimming behaviors: hand-flapping, rocking, jumping, noise making, repetitive behaviors, etc. Sometimes this would increase during the three-week adjustment period and then decrease again. One parent estimated an 80 percent decrease. My family saw eight years of constant head-banging finally end. Most note that any residual stimming is much more mild and better controlled by the child. It is not as 'compulsive.'

Decrease in self-injurious behavior – Some parents report a significant decrease in this type of behavior where it was a problem previously: biting and hitting themselves when angry or stressed, head-banging, throwing one's body onto furniture, the floor, or into a wall.

The Happy Child Effect – This term started soon after establishing the enzyme group. It refers to the general, overall positive disposition of the person once they started taking Peptizyde *and* Zyme Prime regularly. The child (or adult) becomes noticeably more pleasant, easy-going, cooperative, and helpful – just overall very happy. This is in contrast to the fussy, whiny, upset, overall unhappy disposition seen in many individuals before enzymes. Many parents say they were stunned to have their kids consistently get up in the morning in a pleasant mood instead of a grumpy, argumentative one, or just being silent. This effect is seen particularly when Zyme Prime is used regularly. It is the 'happy child' maker.

One mother shared that after her son had been on Peptizyde for one week, and then on Zyme Prime one day, he had been so nice and most of all *quiet* all day. The others in the family kept asking why he was so nice and pleasant today, and *what* did she do? The only thing was the new enzyme. The previous hyperactivity was gone. That little boy continued to be nice, quiet, and happy from then on.

Another mom described her little boy as being 'the best ever' since being back on what they call a normal diet with enzymes. No more red ears, sweaty nights, or black rings under his eyes. But most of all, she was amazed at how happy he is all the time.

At this particular time, we do not know *exactly* why the Happy Child Effect appears, particularly with the Zyme Prime. Remember, it contains enzymes that work across the board for all foods. One thought is that Zyme Prime is used with carbohydrates which may be releasing more nutrients such as magnesium and molybdenum. Magnesium assists with calming. Molybdenum may be necessary in helping to process phenolic chemicals and with sulfation. More trace nutrients may be made available to the body by better digestion of all foods. Better digestion of carbohydrates may be favoring more trytophan, and thus serotonin, to be used. It could help with a hypoglycemic reaction or perhaps gut healing. Whatever is going on, it is a wonderful 'side-effect.' We have seen no correlation between the length of time a child is on enzymes and when or even if the Happy Child Effect emerges. Sometimes it happens right away and sometimes the general disposition improves gradually over time. There is no correlation between the Happy Child Effect and age. An older child (school age) or adult appears to respond as frequently and as quickly as a younger child (preschool). Some people report the Happy Child Effect when they switch to giving enzymes regularly instead of just occasionally. Or in between meals as well as with meals. (P.S. My husband says I represent the Happy Mommy Effect!)

Decrease in pain – Some parents with fibromyalgia, chronic fatigue, migraines, and other painful conditions commented that the pain lessened when taking enzymes regularly. This is common with regular protease use in general.

Improved yeast treatment – Several people said they feel their yeast treatments are more effective while using these enzymes at the same time. This may be because the yeast cells have an outer protein and cellulose coating which the enzymes break down. Once the proteases and cellulases help degrade this protective layer, any medication or yeast-fighters can enter more easily and be more effective. (Note: some enzyme products are available specifically for yeast control.)

Increase in energy – Many adults and children alike on enzymes regularly find they have much more energy than before. This is common with regular enzyme use in general. Like my gaining four extra functional hours each day, or seeing my younger son so much less lethargic. Jordan showed enough energy now to go outside and play, and walk through places himself instead of wanting someone to carry him.

A few people have improved to the extent they no longer fit the criteria for their previous diagnosis. They are no longer on the autism spectrum, or now have a much milder diagnosis, such as language delay; or high rather than low-functioning. My boys were both evaluated as 'recovered' during this time. This was by two different doctors who did not work together, so it represents two individual assessments. They are no longer delayed in any area – functionally, socially, cognitively, or behaviorally.

This is not meant to be taken as a person did nothing else except take enzymes and Poof! he recovered. Almost every family was already doing some type of behavioral therapy, maybe some took medications, or involved in other therapy. Enzymes usually are not the one and only thing needed, but just seem to make a huge difference, making other measures more efficient and effective.

Many families also shared that they saw one of the adjustment reactions that are quite common with most digestive enzymes. These were usually mild, considered quite manageable, and resolved within three weeks. Most of the time positive behaviors occurred at the same time as these adjustment effects.

Although there was overwhelming success, there were those seeing overall negative or inconclusive results, too. Over time these also lessened in frequency as we learned more about enzyme use, digestive problems, and neurological conditions. These are described on pages 336-377.

Of course, with the steady influx of wonderful improvements, we in the enzyme group were thrilled to pieces with these results. We felt officially vindicated with these enzymes proving their staying power. Children were thriving, there were wonderful health improvements, and families were significantly less stressed not to mention financially better off. Enzyme therapy could go forward so others could enjoy these treasures as well.

But yet, we sure wanted to know why all these improvements were happening. Okay, maybe it was just me who was really curious as to what was going on. I like things to make sense; consistency. I mean, I can understand that milk provides calcium and so a nutrient deficiency is resolved, and you might pick up more needed amino acids. Or maybe some peptides are no longer floating around the body. That would provide a boost in improvement. But some of these changes seemed beyond even that. Were *that* many nutritional deficiencies being corrected all at once?

After a little more digging through published research, another few choice gems emerged from the chronicles of enzyme therapy which helped explain all of this.

CHAPTER 13

The Immune System
– Bodyguards on Watch

Our immune system works to screen out any undesirable elements from the environment so they do not enter the body, while allowing in necessary and appropriate substances. Several systems make up the entire immune system in our body. One barrier is the skin. Another is the lining of the respiratory system. The one we will focus on here is the largest one, called the gut-associated lymphatic tissue (GALT). The GALT is intricately linked to the digestive tract. It helps the intestinal lining to block out potentially harmful invaders that come in with food, drink, and air. Current research estimates around 70 to 80 percent of our immune system is located in or around the digestive system. That is a lot of defense being played in the food zone.

The section on leaky gut (Chapter 5) discussed what could happen when the intestines become injured, the intestinal lining loses its integrity, and the selective barrier faulters. This allows food particles, chemicals, and other substances passing through the intestines to cross into the bloodstream when otherwise they would be blocked out. When insufficiently broken-down food enters the body, it may not be recognized as food. It would be nice if the body approached a large compound, saw that it was food needing to be broken down further, and sent it back for more digesting. However, the body

187

registers it as an invader and summons the immune system into action. A case of mistaken identity. Because people eat continually, there is a steady stream of 'invaders' coming into the body when you have a damaged gut. This means the immune system is constantly 'on' at a high level, operating in the mode of flight or fright constantly. The situation takes quite a toll on the system.

The immune connection

Research studies have both identified and measured specific aspects of immune system dysfunction of some sort in many individuals diagnosed with autism conditions (Gupta *et al* 1998; Singh 1998a, 1998b; van Gent, Heijnen, and Treffers 1997).

Jyonouchi, Sun, and Le (2001) tested for immune system activity in 71 children with autism spectrum conditions using various techniques. The results indicated between 49 and 81 percent of the children had excessive immune response, including pro-inflammatory production. Gupta (2002) reviews several studies indicating the immunological abnormalities and significant changes in various immune responses in children with autism. Changes include both deficiencies in some components of the immune system and excesses in others.

So far, it is not known what is the root cause of these abnormal immune responses, just that there is irregular immune system activity (leaky gut? bacteria, yeast, viruses? excessive load to detox? genetic makeup? something else?). This is similar to the compromised immune systems commonly found with other neurological conditions and digestive disorders.

As described earlier, the intestinal tract is over 20 feet long and resembles a fluffy cotton bath towel rolled up into a tube only two inches in diameter. The intestinal tract is lined throughout with a thin, but tough, layer of mucus which covers the stomach, the villi in the small intestine, and the smooth surface in the colon. Within this mucosal layer exists the immune system (along with all those sensitive nerves of the enteric nervous system).

The immune system is filled with an assortment of characters — all having the same job of protecting the body, but having slightly different functions and ways of handling a hazardous situation. They are like the police, the Marines, the air force, fire department, and all the other protection guys. They all work toward the same goal, but come in at different times and for different reasons. As a group, they are all called 'antibodies,' but have other specific names depending on the job they do. Names such as IgG, IgE, T-cells, and B-cells. Any particle identified as potentially harmful or foreign is called an 'antigen.' Antigens are the intruders.

The antibody army of fighters is concealed in the mucosal lining throughout the gut. Clusters of these antibodies reside in the gut lining along with other types of cells that patrol the digestive tract for any suspicious elements. As ground-up food from the stomach flows through the intestines, nutrients are absorbed at particular spots. Throughout the entire intestines, including each villus or microvillus, there is the possibility of intruders gaining entrance to the body.

Whenever something tries to pass the mucosal lining, it is analyzed to see if it should be allowed to pass into the bloodstream, or be sent packing on its way. If something tries to pass that is detected as a potentially harmful invader, the fighting antibodies are called up. This is similar to being screened at an airport security checkpoint. If a foreign substance is detected, specialized cells (like airport security) escort the invader antigen to the lining of the digestive tract (the holding room). There, other specialized cells 'check out' or sample the suspect substance. If they do not check out properly, more special cells are called in to begin processing the antigens appropriately. These other cells take the antigens back to the intestinal mucosa, where they are gobbled up and destroyed by macrophages. Macrophages are like hungry lion cells that welcome a good meal (burp!).

This is how the immune system is so closely tied to the digestive system. Through a complex system of communication, once the immune system identifies an invader, it is able to remember this particular intruder, and be ready for any future attack against a similar substance. This is how immunizations work and why you do not get the same cold virus twice. The first occurrence 'primes' the body so it can recognize that specific intruder and develops elements capable of destroying it. Afterwards, the body is armed, ready, and waiting for another attack. If the same pathogen attacks again, it is destroyed immediately and you do not feel the symptoms and get seriously ill. But that first round may be quite a battle.

Allocation of resources

At any given time, we only have so many antibodies to act as fighting soldiers to protect us against disease. If the antibodies are needed to fight off yeast, there are not as many to protect the rest of the body. If someone has leaky gut and the immune system resources are needed to protect against insufficiently digested particles, then bacteria or viruses are not being fought back as effectively. Also, when the assault on the body is continuous, the immune system starts to wear down, just as soldiers get fatigue and need reinforcements. You may be sending nourishments down to supplement your own body and immune system, but if yeast/bacteria intercept the supply line, they will use it to grow instead while the immune system weakens.

Other reserves and troops can be called up but this takes time as well as resources and energy from the body to manufacture and deploy them. Just as when a country goes to war, the more resources it has to put into the war effort the less resources are available to go toward building a healthy country and economy. Roads and schools become secondary concerns (maintenance of the body is slowed). New buildings are postponed (children's growth wanes). Citizens have to make due with rations and substitutes (nutrient deficiencies, toxins build up). If the body cannot get the minerals for certain processes from food or supplements, it may start scavenging the raw materials it needs from bone and muscle.

And eventually, just as troops may lack proper fighting weapons, food, and supplies to keep it going, the immune system will weaken and not have the resources it needs to defend the body as it should. This is a fundamental reason that people who get one pervasive condition, such as yeast overgrowth, leaky gut, or digestive problems, start to accumulate even more problems as the entire body begins malfunctioning.

Think of trying to walk up an escalator going down. You have to put in effort just to break even. By standing still, you naturally drift down. Similarly, until some of the fundamental problems are addressed, your health may continue to gravitate downward. The good news is that healing can work in the same way. A positive effect on one part of the system has a positive effect on other parts. An important aspect of digestive enzyme therapy is it works much closer to the root of the problem, and on so many levels.

How enzymes assist the immune system

So how exactly does supplementing digestive enzymes help strengthen the immune system? Enzymes help heal a leaky gut by breaking down irritating food particles and cleaning out debris. Enzymes break down food so that anything passing into the bloodstream will be less problematic. Enzymes enable the basic components from the food to be available in the form of raw materials the body needs. They also prevent the arteries from clogging and joints from becoming gummed up. They may clear out the damaging free radicals. All of these lessen the strain on the immune system. Enzymes also increase white blood cell size and activity, and raise T-cell activity and production (both are some of the agents of the immune system), and act as immune system regulators (Bonney and Davies 1980; Lehmann et al 1992).

With poor digestion, the immune system may have many more things to act on (invoking the IgE allergy or IgG intolerance reactions). This results in two undesirable events. You do not get the nutritional and energy benefit of the food, and your immune system is called into action. Your immune system now needs to spend time and energy cleaning up the bloodstream. While they are busy doing this, they are not at full strength to protect and repair the body. With the body all tied up combating an onslaught of invaders, very few resources or energy are left to do other things. Important things such as building a healthy body, learning, working, having conversations, playing, and enjoying a good quality of life!

Whether low enzyme production has a genetic basis or results from some other cause, low enzyme production is known to have a devastating effect on overall health. Many individuals with cancer, diabetes, heart disease, arthritis, and other degenerative conditions show low or deficient levels of enzymes. Whether the lack of enzymes cause the disease state or simply is a result from a particular condition will depend on the illness itself.

What is more interesting is the observation that resupplying digestive enzymes to people with such degenerative conditions has rapidly improved their health. Many clinical studies proving enzymes can slow or even reverse these disease states are available. This is the part I like – once you find yourself with a certain problem, what can you *do* about it.

Autoimmune conditions and enzymes

One view of the biological aspects of autism, and perhaps even extending to AD(H)D, looks at symptoms being based in a neuro-immune system dysfunction, or even being an autoimmune condition. Some organizations and research projects focus just on this issue. One explanation is from Vijendra K. Singh, Ph.D., *Neuroimmune Biology, Volume 1: New Foundation of Biology.*

> Based on our ongoing research of a reciprocal relationship between nervous system and immune system, we studied autism as a neuro-immune dysfunction syndrome in which autoimmunity to brain was strongly implicated... We found that auto-antibodies to MBP (myelin basic protein) were selectively present in up to 80 percent of the autistic children, but they were only rarely detected in the controls...The idea that autism may be an autoimmune disorder [in some] is further strengthened by the fact that many autistic children respond well to treatment with immune modulating [treatments]... In conclusion, autism involves a neuro-autoimmune response that occurs at the neuro-immune biology interface. Clinically, therefore, there is enormous potential for restoring brain function in autistic people through immunology.

Myelin is the substance that coats the nerves insuring proper growth and function. Having auto-antibodies to this indicates that something in the body is attacking its own nerve structure.

It is amazing the number of either one or both parents of children with neurological conditions who also have some sort of compromised or dysfunctional immune system. These seem to run in families and support a genetic contribution to the conditions:

- allergies
- autism/Asperger's /PDD
- Alzheimer's disease
- cancer
- celiac disease
- chronic fatigue
- colitis
- Crohn's disease
- dermatitis herpetiformis
- inflammatory bowel disease
- asthma
- AD(H)D
- fibromyalgia
- lupus
- Sjogen's syndrome
- thyroid diseases
- psoriasis
- multiple sclerosis
- rheumatoid arthritis
- type 1 diabetes

You might also express it as a genetic predisposition to immune system related conditions. Which symptoms actually appear may depend on a combination of the individual's environment, experiences, and biochemistry, and which major organs or parts of the body are affected. In our family, there is a connection so blatant it shines like a full moon. That I have long had a neurological 'problem' is an obvious marker. However, the susceptibility of our boys may have also been foreshadowed by my husband's life-long asthma and irritable bowel syndrome. Or my grandmother's and great-grandmother's crippling arthritis. Maybe some folks just have not thought of the possible correlations. One of your uncles had bipolar, or a cousin was allergic to 'everything,' or an older grandparent coming down with Alzheimer's. Or just an assortment of relatives with a mixture of seemingly unrelated aches and pains.

No one is completely sure what causes particular autoimmune diseases although a mixture of genetics, environment, and maybe pathogens is involved. Currently, no cures for most autoimmune diseases exist. However, many good treatments are available that allow you to live around the condition, have a good life, and prevent the situation from worsening. Some of the targeted therapies aim to keep the immune system working in tip-top shape. A two-sided approach involves 1) keeping the immune system strong by adding supplements, medications, and therapy, and 2) avoiding things that stress it.

A key source of problems with autoimmune conditions is inflammation. Inflammation is the cause of much of the pain associated with these conditions. If inflammation persists, tissue is destroyed.

Managing and controlling inflammation goes a long way towards limiting the damage caused by an autoimmune condition. Remember how proteases are so effective for controlling inflammation? (refer to Enzymes and Disease, Chapter 8)

One of the problems with medications commonly used for autoimmune conditions is they suppress the immune system. Suppressing the immune system reduces the level and intensity of attack your body is waging on itself. You are decreasing the size of the army and fire power. The idea is to slow down the rate of damage. However, slowing or suppressing the immune system will also prevent the body from its natural ability to protect itself from invaders. Now you are exposed to developing other illnesses.

This is one of the major blessings of using a natural substance such as digestive enzymes. Enzymes are very effective at building up the immune system while decreasing inflammation, all with none of the side-effects of many medications. Fortunately, we do not have to guess at how well enzymes may help autoimmune conditions. Enzyme therapy has an excellent track record in this area.

In his book *Enzymes and Enzyme Therapy*, Dr Anthony Cichoke outlines the research conducted on how enzymes work to eliminate the all-important circulating immune complexes that provoke so many autoimmune diseases. Without getting too technical, basically when something enters the body that is seen as potentially harmful, it is called an antigen. The body then produces things called antibodies, which have the specific function of binding to the invader to render it harmless (like a policeman handcuffing himself to a crook so he can't run off). This antigen plus antibody group is called an 'immune complex' or IC.

Under normal healthy conditions, these complexes are eliminated from the body immediately. However, in other situations, the body may be overloaded and not be able to clear out the complexes (like a jail that is already full). The immune complex may continue traveling on in the bloodstream. At this point, it is called a circulating immune complex, or CIC. These complexes may come to rest in, or become trapped in, a number of possible sites in the body. They are deposited in such places as the lungs, kidneys, skin, joints, or blood vessels.

When this happens, the complexes can cause inflammation and tissue damage. This is the making of an autoimmune disease. Which condition you may have will depend on which place in the body the complexes are deposited. Some CICs cause rather creative adverse reactions, while other CICs may increase suppressor effects. Some may do both, or perhaps something else entirely.

Our bodies contain a series of checks and balances to ensure that our immune systems do not attack the 'good' necessary tissues and organs. Each cell in the body carries a few molecules that identify it as being an appropriate part of the body – like carrying an identification badge. The immune system sees this badge and lets the cell go on about its business. Cells not having the appropriate markers set off the alarm, and the immune system springs into action. Our bodies have a very complex and sophisticated communication system and method of identifying and attacking intruders so the body is well protected against its environment, yet allows its own biological processes to function normally.

However, at times this system breaks down. For reasons not understood, the immune system may see some of its own cells as invaders. It creates a series of antibodies to attack and tie up some of the body's own cells. This also leads to more immune complexes, which may end up circulating and becoming lodged in the body. Testing shows many people with autoimmune diseases have a much higher number of these immune complexes, either lodged in tissue, in circulation, or both.

Will getting rid of these complexes help? Yes, very much. Research since the 1970s has shown that eliminating these immune complexes improves many conditions including:

- Crohn's disease   (Fiasse *et al* 1978; Kre *et al* 1980)
- hepatitis   (Musca *et al* 1984; Stauder 1989)
- HIV infections   (Stauder *et al* 1989; Hager 1990)
- multiple sclerosis   (Dasgupta *et al* 1982; Kidd 2001)
- pancreatitis   (Theofilopoulos 1980)
- rheumatoid arthritis   (Fehr 1984; Klein *et al* 1988)
- ulcerative colitis   (Hodgson *et al* 1977)
- various tumors   (Sakalova *et al* 1992; Hellstrom *et al* 1985)

. . . and a number of other immune system disorders

If we know that getting rid of these complexes helps, how can we eliminate them? Certain mixtures of hydrolytic enzymes, including proteases, lipases, and amylases have reduced the number of circulating immune complexes in past studies (Stauder 1990; Stauder *et al* 1989; Ransberger *et al* 1988; Targoni, Tary-Lehmann, and Lehmann 1999). These enzyme mixtures have brought significant improvement to the patients. Enzymes work by breaking up the harmful complexes and activating the macrophages, which gobble up and destroy the intruders. This ends the vicious cycle that leads to deterioration and many chronic disorders.

In 1983, Dr Steffen of the Institute for Immunology at the University of Vienna treated rabbits suffering pathogenic immune complexes with a concentrated enzyme solution. In the end, he found that the more concentrated the enzyme mixture was, the more immune complexes were destroyed. Eventually, all of the immune complexes were broken down within a few hours and the rabbits returned to good health (Steffen and Menzel 1983).

In 1985, Dr Steffen *el al* treated 42 patients with definite or classic rheumatoid arthritis with an enzyme blend for six weeks. The amount of circulating immune complexes during treatment were measured. Results indicated that 26 (61.9%) patients improved as the immune complexes were reduced. No side effects were observed.

Drs Ransberger and van Schaik studied the effect of digestive enzymes in patients with multiple sclerosis (1986), a disorder where immune-complexes play a prominent role in this neuro-autoimmune condition where the myelin coating is disrupted or destroyed. Very good results were noted with the digestive enzymes reducing severity of symptoms overall, including visual alterations and disturbances, sensory dysfunctions, and intestinal problems.

In 1986, Dr Mertin and Stauder reported on a group of 300 multiple sclerosis patients treated with hydrolytic enzymes. The results indicated the enzymes stabilized the disease, reduced the relapse rate, stabilized the neurological impairment, and improved daily living.

Dr Neuhofer is one of the best known physicians treating multiple sclerosis with enzymes, treating hundreds of cases. In 1986, 83 fully evaluated multiple sclerosis patients used enzyme therapy with 85.4

percent improving: 45 showed substantial improvement (54.2 percent) and 26 showed some improvements (31.3 percent). All of the non-responders had been on long-term immuno-suppressant medications.

Another area that enzymes assist with is the elimination of free radicals. Free radicals are the natural by-products of our regular metabolism. Free radicals are highly reactive, electrically charged molecules that can react with certain cells in the body and disrupt biological functions. They are capable of causing cells to lose their structure and function, and can eventually destroy them. They can damage tissues of the nervous system, joints, internal organs, and blood vessels. The general term for free radical attack is oxidative stress. Toxins and pollution from the environment can also contribute to the free radical load.

Besides using enzymes, free radicals can be controlled by beneficial compounds known as antioxidants. Antioxidants are capable of stabilizing or deactivation free radicals before they attack cells, or they can even stop the free radical from forming in the first place.

Antioxidants are a classification of several substances, and often involve phenolic compounds. Common sources of antioxidants are:

- fruits and vegetables, particularly dark-colored ones
- vitamins C, E, and vitamin A (the 'A' from beta-carotene)
- carotenoids (including beta-carotene; give the color to many fruits and vegetables)
- selenium (a mineral)

Over 60 diseases including such autoimmune diseases as chronic fatigue, fibromyalgia, and cancer involve free radicals (Ames 1993).

Cancer treatment with enzymes

Many different types of cancer have been identified. Cancer has been treated with enzymes for many decades. Dr John Beard found that pancreatic enzymes could be used to successfully halt out-of-control cancer cell growth. His patients lived longer and cancer growth was inhibited with enzymes. The book *The Enzyme Treatment of Cancer and Its Scientific Basis* contains his findings of treating 170 patients and is considered a fundamental cornerstone of all enzyme therapy practiced today for cancer (Beard 1911).

Dr Max Wolf treated about 50,000 cancer patients with various enzyme combinations over a period of 25 years. That is a lot of people over a significant span of time. He found that certain combinations had a synergistic, and therefore, more effective, action as opposed to using individual enzymes. This is the 'art' of enzyme formulation! The same way the particular combination of enzymes in Peptizyde worked differently than any previous combinations of proteases. Most cancer patients live longer and healthier lives when taking enzymes, establishing a solid place for digestive enzymes in cancer research and treatment.

This beneficial effect of enzymes in cancer is related to several logical factors. First, cancer cells are more susceptible to proteases than regular cells. Cancer cells are covered by a protein film that proteases break down. This leaves the cancer cell open to the body's natural defenses, chemotherapy, or other measures.

Enzymes also stimulate the body's natural cancer fighting weapons. They help disarm the cancer cells. Next, enzymes can break down the sticky 'glue' substance that cancer cells use to stick themselves onto other cells and tissue. This retards their spread and growth (Desser 1990; Gonzalez 1999). Should someone need surgery, radiation and/or chemotherapy, enzymes can reduce the side-effects that usually accompany these treatments. Enzymes help reduce the amount of inflammation and pain experienced, and speed up the healing process (Ottokar 1980, Wrba 1990). Experience shows enzyme therapy is effective with many types of cancers including those of the skin, digestive system, connective tissues, breast, lymphoma, leukemia, and others. (Gerard 1972; Desser, Rehberger and Paukovits 1994; Lauer et al 2001; Leipner and Saller 2000; Nieper 1974; Taussig, Szekerczes, and Batkin 1985.)

Arthritis

Arthritis and other joint diseases are marked by inflammation and the presence of immune complexes. Because enzymes are very effective on both these factors, they brought a new light to the world of this painful and debilitating autoimmune condition. In 1983, a study was published on the use of enzyme treatment with 1004 rheumatic patients and involved 141 doctors. (Did you catch that number of patients and doctors?!) The patients had a wide variety of conditions including arthritis, soft tissue rheumatism, rheumatism of the joints, and other rheumatic disabilities. From 76 to 96 percent were 'improved' or 'considerably improved' with enzyme therapy and 1 to 24 percent remained 'unchanged' depending on their ailment. Over 99 percent reported they were free of any side-effects with enzyme therapy (cited in Glenk and Neu 1990).

Compare these results with the first summary from our enzyme group which showed 87 percent of those trying the new enzymes reported improvement, which agrees nicely with the 76 to 96 percent range found here.

Dr Klaus Uffelmann of Genundsen, Germany reported on an eight-week, double-blind study with 424 arthritis patients. Protease enzymes proved particularly effective in alleviating pain, improving mobility, and reducing soft tissue swelling and muscle stiffness. These results continued past the eighth week even after discontinuing the enzyme doses (Uffelmann 1990; Cichoke 2000). For over 20 years in Germany, doctors have continued to successfully treat osteoarthritis with enzymes, and consider enzymes equivalent in effectiveness to medications (e.g. Singer 2000; Nouza 1994).

The eye of the beholder

In a paper published in 1995, Stauder reviews the following pharmacological effects of combinations of oral digestive enzymes:

- fibrinolytic (clot breakdown or removal; improves circulation)
- hemorheologic (helps blood flow; improve transport of nutrients and other compounds)
- anti-edematous (reduces water retention and swelling)
- anti-inflammatory (reduces inflammation and pain)
- activation of macrophages and Natural Killer cells (NK) (strengthens immune system)
- modulation of adhesion molecules, cytokines, and immune complexes (regulates and strenghtens immune system)

The pharmacological effects may be classified on four levels:

- biochemical
- immunological
- medical
- physiological

These are the extensive, well-documented health benefits digestive enzymes are known to provide besides food breakdown and nutrient release. They also go a long way toward explaining the amazing improvements we were experiencing using enzymes.

Most enzyme blends used for particular illnesses such as multiple sclerosis, cancer, arthritis, etc. are particular mixtures with the individual's needs and history taken into account.

The idea behind using enzyme therapy to correct the autoimmune function is that you are not just treating symptoms, you are helping to re-establish the correct regulation and support of the immune system. Enzymes tend to help calm the immune and neurological systems down. Gently . . .softly . . .and very smoothly. Calming them down, reducing the stress load, regulating them appropriately.

If enzymes are so very beneficial with immune system disorders, why not in other conditions having an immune system connection? Since many people with autism spectrum conditions suffer some type of immune system dysfunction, it is very reasonable to expect enzyme therapy to be effective. Treating the biological problems of any neurologically based condition with enzymes may be a new idea, but there is substantial, proven, science-based evidence that enzymes are a logical and very effective therapy.

And that is exactly what we saw happen – immense improvement for most individuals. The remarkable progress was not imagined, conjured up through wishful dreaming, based on a mysterious quantum leap of faith, or a random fluke. It shows what we are experiencing is grounded in sound principles of fundamental science and physiology. This is another direct, well-founded explanation why parents were seeing such astounding 'unexplainable' improvements in their children.

And there were a few more to come.

CHAPTER 14

Dysbiosis

– Life in the Gut

My younger son, Jordan, has always had a terrible time with gastrointestinal issues. He would alternate between diarrhea and constipation. Stools were foul smelling. We are talking Stench – the kind of smell that could cause a stampeding elephant to drop in its tracks. The kind you can smell over the telephone. He was a mess, literally, through the age of seven. It's not as if I didn't take him to the doctor and try to find out what to do.

'He'll grow out of it.'

'He's doing it for attention.'

'He needs to see a psychologist for behavior therapy.'

We went to behavior therapy for two years. After exhausting all the bowel-training suggestions known to mankind, and burning through all the possible reasons he might be 'doing it as a power struggle,' I took him eight different times to a medical doctor just for this issue. I listened to all the standard reasons for the problem again and waited for Jordan to 'grow out of it.'

But he did not grow out of it. Finally, we came away with 'He probably has irritable bowel syndrome and there isn't much you can do about that anyway. Good-bye.' Except this doctor also said that

Zoloft and amitryptiline were often prescribed for and helped irritable bowel syndrome, although they were not sure why. Maybe because of the sensory issue there. We were already taking those medications, but it was, yet again, interesting that this all seemed related somehow.

Jordan complained of headaches and bounced his head a lot. Since an infant, he disliked sunlight in his eyes or loud noises. He *always* wanted someone to hold him or carry him. And always clutched Little Bear while twirling that green ribbon. He did not like to sleep alone or without lights on. His sensory issues were not as debilitating as Matthew's, but were definitely there. Maybe they were actually a lot worse, but since he was always so quiet, we could not tell. I suspected he just expressed his discomfort differently instead of raising a ruckus. He did not seem to get quite the relief that Matthew and I had from the medications, or being dairy-free.

Jordan was the p-i-c-k-i-e-s-t eater and ate incredibly slow. He would take a good 60 to 90 minutes to eat a small lunch. He whined for a selection of only five or six foods (cheese or related dairy products, bread products, chicken nuggets, fish sticks, cheeseburgers, and French fries). I insisted he eat vegetables and fruit, but it was a struggle.

On top of that, he tended to gag up randomly. It was never over something specific, he just appeared to have a very sensitive stomach or reflux. His dad commented that he was looking rather anorexic – the sickly thin, pale look with dark rings under his eyes. He was not gaining weight and was always very, very tired and lethargic. Jordan's eating became a major issue at our house.

The first two weeks he was on enzymes, he did not show that much change. However, in the third week, we were in for a major shock. Jordan came to me, begging for food. He said, 'Mommy, I am S-T-A-R-V-I-N-G! Pleeeease can I have something to eat!'

I was surprised, but this was a wonderful turn of events. He became ravenous over night. He jumped up to eating about one-third more food each day, ate in a reasonable 20 minutes, and quickly began accepting around 30 different foods, including fruits and vegetables (but not lima beans – however, I can't stand lima beans either; the only good thing about lima beans is that they are not orange).

In the first three months, he put on five pounds he desperately needed. Then he continued to gain three more in the next two months. He now looks healthier and is much more active, and much more alert. He has not gagged up in quite a long time either.

When we started Peptizyde, he had the loose stools for two days and then got *much* better. Enzymes at every meal eliminated about half of the horrible smell. In addition, we started giving him Culturelle probiotics (beneficial microorganisms) three times a day, which seemed to help. I gave him extra Peptizyde for the proteases in between meals to help clear out waste and toxins in the colon. I did not know it at the time, but because the constipation had been so bad for so long, he had developed both encopresis and a terrible bacteria overgrowth, hence the smell. Both of those needed to be treated separately now. One problem leads to another problem and then another.

After five months of giving an aggressive amount of enzymes and Culturelle, the bacteria problem was kept manageable, but not conquered. Each time I reduced either the enzymes or probiotics, the smell came back just as strong as ever. Based on assistance and advice from other parents, I figured there was some sort of pathogen problem. Either yeast, bacteria, or parasites. Because testing for pathogens is not always reliable and there was a bit of expense and hassle to it, I decided to systematically treat for each type of pathogen with over-the-counter remedies and see if anything improved. Then, I could always look into the appropriate prescription treatment, if necessary.

First, I considered yeast overgrowth because that was very common with gastrointestinal problems in the general population, and often with autism conditions. I bought some grapefruit seed extract, which is surprisingly effective for yeast control, from the health food store and gave it to Jordan for a few days. Nothing. I was looking for a negative reaction that would indicate die-off. A negative reaction would have happened right away. Nothing happened for two weeks. So then, I bought an herbal blend for parasites and gave that for two weeks. Nothing. So then, I looked into using a natural antibiotic to treat bacteria overgrowth. A health practitioner can advise appropriate items for various pathogens.

BINGO! The first day on the antibiotic Jordan felt quite sick to his stomach and had a particularly bad headache. This continued through the third day. Then he felt much, much better. I continued giving it for ten days just as you would any antibiotic. We had brilliant success! Finally, the horrid smell was gone; and stayed gone for months afterwards. After thinking about it, it makes sense that since Jordan had such a bad colon problem that this would create a bacteria overgrowth. He has done better ever since then. His gut is finally functioning well. My original three wishes granted.

There goes the neighborhood – The science behind it

The gastrointestinal tract is a very dynamic place just budding with a variety of life. Trillions upon trillions of bacteria inhabit our colons breaking down all the waste passing through. An assortment of characters live in our gut, in a delicate balance with each other. With an appropriate balance, bacteria and other microbes play an important and useful role in digestion and contribute to overall health. The beneficial ones, the good guys, are termed probiotics (meaning pro-life) because they are friendly to life by promoting our health and well-being. Probiotics improve the environment of the intestinal tract, and are important in healing many chronic gastrointestinal problems.

The bad guys are the unfriendly creatures that disrupt proper digestion, churn out toxins, throw our systems out of balance, and cause us to become sick in a multitude of ways. They may be acquired from the environment, or grow internally. Conditions can arise that may kill off the 'good' microorganisms and allow the 'bad' microbes to flourish and take over. When the adverse microbes take over, every type of havoc is created in the gut, which spills over and affects the rest of the body.

Dysbiosis is a general term that refers to any imbalance in the normal healthy diversity of the many microbes that inhabit our intestines. Dysbiosis can develop at any stage of life. However, certain conditions seem to bring it on. The use of antibiotics (meaning against life) often results in killing off the friendly elements as well as the harmful ones. And like weeds, the problematic bacteria and microorganisms can survive in more acidic and unpleasant environments than the beneficial guys. In the beginning of antibiotic use, probiotics were a regular part of the treatment, being prescribed to replenish the supply of helpful bacteria. But somewhere along the way, the probiotics part got lost. It is a good idea to take probiotics whenever antibiotics are used.

Constipation can throw off the microbial balance in the gut by keeping the waste and toxins in the colon longer. The potentially harmful bacteria thrive better in such a polluted environment than the beneficial organisms. Be prepared for harmful bacteria overgrowth if someone suffers impaction or encopresis, the more severe forms of constipation.

Any type of digestive problems as well as chronic illnesses can also cause the balance to be thrown off. Dysbiosis can promote:

- gut inflammation
- leaky gut
- food and chemical allergies/intolerances/sensitivities
- maldigestion and malabsorption
- constipation, diarrhea
- liver toxicity
- immune system dysfunction
- toxins disrupting neurological and metabolic pathways

…and all the problems these things bring with them. This list is starting to sound very familiar, isn't it?

Dysbiosis is common enough in the general population, but seems particularly abundant in the autism spectrum. Perhaps it is because many with autism or related conditions have systemic dysfunctions, such as suppressed immune systems or insufficient digestion, that allow the harmful microorganisms to take over. But equally likely is that the neurological symptoms emerge due to the effects of the many neuro-toxins produced by the not-so-beneficial microbes.

Notice that young children often 'become' autistic or 'develop' problems as toddlers, which is also the time many children start having ear infections and antibiotic treatments (Konstantarea and Homatidis 1997). I can remember Matthew at that age chugging down so many rounds of antibiotics for ear infections, I might as well have put it in his bottle. There may be a group of children that actually suffer from an unrelenting harmful yeast, bacteria, or other pathogen problem which might be overlooked. When the dysbiosis is treated, they 'recover' from their previous problematic behavior.

The not-so-helpful bacteria

Several strains of harmful bacteria in particular seem to spring up. *Clostridia* and *Klebsiella* are common. *Clostridium* produces a potent neuro-toxin that affects the central nervous system that could cause a wide variety of behavioral problems. Research by Sandler *et al* (2000) verifies that the neuro-toxins from bacteria can contribute to or cause the symptoms expressed in autism based on seeing a reduction in these symptoms after treatment with antibiotics.

Adverse bacteria can give off excess ammonia as well. If you smell ammonia, or have test results indicating high ammonia, consider a bad bacteria overgrowth. High ammonia is very harmful if it builds up in the body. Certain types of enzymes are used frequently in animal science to cut down on ammonia output. The idea is to give enzymes that can break down the hard to digest plant fiber and roughage so there is little left over to feed adverse bacteria. These include enzymes such as xylanase and cellulase. Any insufficiently digested food reaching the colon is simply feed and fodder for the bacteria there, and has a tendency to increase the bad gut bugs. Adverse bacteria overgrowth is a common feature in many colon problems. More on this and bacteria in Chapter 15.

In our case, although the enzymes helped control the harmful bacteria problem, they did not conquer our severe problem. Ultimately, we needed to treat specifically for the overgrowth. I think the enzymes worked very synergistically with the bacteria treatment making it much more effective than not giving enzymes at the same time. I later found research reaching back to 1961 supporting my observation.

In a study titled 'A plant protease for potentiation of and possible replacement of antibiotics,' Neubauer found that twenty-two out of twenty-three people who had previously not responded to antibiotics did so after adding bromelain four times per day. Also, bromelain improved the action antibiotic medicines, including penicillin and erythromycin, in treating a variety of infections. Later, Tinozzi and Venegoni (1978) showed that bromelain increased the absorption of the antibiotic amoxicillan in humans. Enzymes have improved overall recovery when added to antibiotic therapies in other bacteria related illnesses including ear infections and systemic infections (Ivaniyta *et al* 1998; Veldoza *et al* 1999).

The definitely dastardly yeast (*Candida*)

Yeast is often a fellow inhabitant of our intestines and mucosal system, including the skin, lungs, and throat. At times yeast grows out of control which can throw the ecological balance of the gut into chaos. Yeast is not necessarily confined to the gut. It may migrate and invade other locations in the body leading to pervasive dysfunctions.

Research finds yeast overgrowth to be related to autism, AD(H)D, psychosis, skin problems, thrush, vaginitis, athlete's foot, chronic ear infections, hypoglycemia, food intolerances, malabsorption, chemical sensitivities, depression, ulcer, inflammation, and autoimmune conditions such as fibromyalgia, multiple sclerosis, and chronic fatigue syndrome. Yeast can produce an array of hard-to-pinpoint symptoms including migraines, headaches, moodiness, hyperactivity, social withdrawal, clumsiness, stomach aches, and bowel problems; as well as be the basis of sensory dysfunctions. (Our all-too-familiar list of interrelated syndromes and symptoms.)

While anyone can have these symptoms sometimes, the difference is when these occur inappropriately and to a debilitating level. When it is chronic and uncontrollable, not otherwise explainable. Many people in the general population are finding they have a challenging yeast overgrowth to deal with. Our little ones are just as susceptible.

A fair amount of children end up with a yeast problem. Based on trends seen in testing labs, it may be as common as 70 percent of children with autism conditions having some degree of yeast overgrowth. This is thought to be caused at least partly by the ever-increasing use of antibiotics, especially for young children's ear infections. Antibiotics favor yeast. Antibiotics knock out all the bacteria, but do not affect yeast. So the yeast gets plenty of room and nourishment to multiply, grow, and stretch throughout the gut. Diets high in sugar, refined carbohydrates, and processed foods also stimulate yeast growth – very characteristic of young children's diets.

A particularly problematic strain of yeast is *Candida albicans*. Yeast is a fungus that can exist in two forms. The single-cell stage is considered the form that is generally not problematic. However, yeast can grow out of control and evolve into its fungal form. The fungal form is especially hazardous to the body for several reasons.

First, it grows rootlike structures that snake out from the core. These hyphae bore into the mucosal lining in the intestines. Like little extended earthworms, they plow through the protective layer of the gut. This allows any number of substances to seep through the intestines and into the bloodstream. Yeast overgrowth is a primary cause of leaky gut syndrome.

Another problem with yeast overgrowth is that the yeast give off a fair number of toxic by-products as they grow and when they die. Many of these substances are neuro-toxins, very harmful to the nervous system as well as the brain. Among the numerous toxins identified are carbon monoxide, alcohol, and acetaldehyde. Acetaldehyde is particularly volatile and toxic, potentially causing a multitude of neurological, endocrine, metabolic, and emotional problems (Truss 1984). Alcohol is a by-product produced by yeast and gives rise to the 'drunk' behavior seen in some people (Rochlitz 2000 Chapter 7). Inappropriate and hysterical giggling or laughing, and night waking are common symptoms as well. These symptoms can occur either very rapidly after eating or appear later. The neuro-toxins greatly disrupt sleep, another important reason to get it under control. Dealing with all of these neuro-toxins at once would create 'unique' behavior in anyone, without the additional factor of a young child's developing brain, nerve, and immune systems.

Yeast can do another little trick. Yeast converts inorganic mercury, which may enter the body by a variety of sources, into methyl mercury, a very quick-traveling toxic form that readily penetrates tissues throughout the body, including the brain. What type of damage occurs depends on the organs the mercury settles in. In addition, mercury and other metals can suppress the immune system, creating a nice smooth path for yeast or other pathogen to take off.

As the yeast grows, it keeps constant pressure on the immune system to fight it back. If the immune system is already taxed fighting other stressors, it might not have the resources to continually keep the yeast in check. Or if the immune system is kept busy controlling yeast, this leaves an opening for other stressors to get a foothold in the body. Either way, yeast is major trouble and biological wellness heads in a downward spiral. The extensive immunological impairment possible just exasperates immune system dysfunctions.

While some dispute the extensiveness of the yeast problem and feel it is greatly over-hyped, other studies describe the possible tissue injury, emotional/mental disturbances, and neurological problems associated with yeast and dysbiosis in both young children and adults (*The Missing Diagnosis* Truss 1983).

All this growing and manufacturing of toxins requires a lot of input. Where do you suppose the yeast get all the energy and nutrients they need to grow? You guessed it. Yeast are busy gobbling up all the food and supplements you send down intended to nourish your own body. The yeast grow nice and strong while you incur nutrient deficiencies.

Many children show significant improvement in their 'autistic' behavior once the yeast overgrowth is treated. Dr William Shaw from the Great Plains Laboratory in Lenexa, Kansas has conducted notable research in this area. His information reports that of more than 1000 children with autism conditions, a good clinical response was shown (around 80 percent or better) when these children were treated with a wide variety of antifungal agents such as Nystatin, Lamisil, Sporanox, Nizoral, Diflucan, caprylic acid, grapefruit seed extract, and garlic extract. It is not clear if all these patients had identified yeast to begin with or not, or how severe the cases were. Parents ranked antifungals as one of the most effective drug therapies used for the reduction of autism symptoms. Yeast treatment has resulted in decreased hyperactivity, better eye contact, increased language, improved sleep patterns, improved concentration, increased imaginative play, reduced stereotypical behaviors, and better academic performance (see *www.greatplainslaboratory.com/yeast.html#facts*).

Some tests for yeast are available, but they are usually not 100 percent reliable or conclusive. Yeast overgrowth is usually diagnosed based on history, numerous symptoms, and confirmed by a positive response to yeast treatment. A yeast treatment usually involves prescription or over-the-counter antifungals (yeast killers), a yeast control diet, support supplements, and avoiding other stress on the immune system. Prescription medications like Nystatin or Vancomycin often help. Over-the-counter remedies include grapefruit seed extract, cranberry extract, and other herbs.

A special diet usually includes avoiding sugar in all forms (corn syrup, honey, fructose, barley malt, etc.), refined white flour, and processed foods. Fruits including juices are avoided or limited, as are bakery products and other foods containing molds or yeast. This includes ketchup, mushrooms, cheese, vinegar, and fermented beverages. A yeast diet focuses on meats, and complex and low carbohydrate foods. Supporting nutrients include vitamin C and magnesium. The Further References section at the end of this book gives sources for various yeast diets, symptoms, and how yeast contributes to neurology and biological functions.

Treating yeast takes some time. Treatment focuses on strengthening the entire immune system, getting the gut micro-organisms back in balance, and restoring proper digestion. This is why digestive enzymes become such an important member of the yeast-treatment team. They excel in supporting all three aspects of treatment.

We have already discussed how enzymes are so effective in strengthening the immune system as well as instrumental in gut healing. Certain types of enzymes also happen to be top-rate yeast fighters. Yeast cells have tough outer shells made out of protein and cellulose. Proteases and cellulases are two enzymes that can break down these tough outer shells. This allows the antifungals to gain access into the yeast cells more readily, killing them off more effectively. Proteases then assist with carrying off the debris.

When yeast dies, it gives off an assortment of toxins, potentially creating a great deal of adverse reactions. The neuro-toxins affect the nervous system in particular. Enzymes assist in carrying off the toxins and helping the body recover during this time of die-off.

Several parents using a prescription yeast treatment without enzymes and then with enzymes reported the treatment was noticeably more effective with enzymes. They also vouched there were far less negative die-off reactions. Other parents reported they were unable to completely subdue a yeast problem until enzymes were added. This was consistent with our enzymes-plus-antibiotic experience.

The die-off reaction can take anywhere from a few days to a few weeks depending on how aggressively you are treating the yeast and how bad the yeast overgrowth is. *The Complete Candida Yeast Guidebook* contains a list of many of the wide range of possible die-off symptoms and suggestions for treating each one. One note: it is advisable to monitor the child's progress or your own carefully during the die-off phase. If the person seems too uncomfortable and the die-off is more severe than you are comfortable with, treat more slowly. This is one case where trying to 'tough it out' hoping to get through it faster may not be the best route. The reason is that the immune system may be stressed so much by the toxins released from the die-off it is not adequately able to fight off the yeast. A weakened system may only allow the yeast to grow more swiftly again, perhaps in another area of the body. So you may not be gaining much ground overall.

Drinking as much pure water as possible and encouraging proper bowel elimination, even to the point of very loose stools, helps wash the toxins out quicker relieving symptoms. Activated charcoal and bentonite clay may be effective in soaking up and removing the toxins.

Some enzyme products are designed just for fighting yeast. Enzymedica makes a product that contains just proteases and cellulases called Candidase. Candex from Pure Essence contains a large amount of cellulase. Garden of Life makes a whole food product including proteases and cellulase along with some of the antifungal herbs given previously. Strong protease products like Peptizyde may help with yeast and other pathogens. No-Fenol from Houston Nutraceuticals contains a fair amount of cellulase with other assisting fiber digesting enzymes. You may want to try taking different enzyme products together to achieve a combination that will meet your needs. Candex or No-Fenol plus Peptizyde could be a strong combination for yeast. A large number of people have reported brilliant success fighting yeast with No-Fenol plus grapefruit seed extract.

Enzymes have a long association with yeast. The word 'enzyme' is derived from greek and means 'in yeast' due to the study of glucose being converted to alcohol by yeast cells. Before that, this action was attributed to some mysterious life force energy.

The problematic parasites

Fungi, bacteria, and parasites are made up of protein. Protease enzymes can break these critters down, thus destroying them. Taking proteases between meals can help rid the body of these unwelcome invaders, and thus help the immune system. Pancreatic insufficiency also favors parasites because the digestive enzymes that should be present from the pancreas destroy the parasites as they enter the small intestine. Another situation where supplementing with oral enzymes may help.

The stomach acid usually kills the parasites. If someone has low stomach acid production, the individual is more likely to acquire parasites. Parasites may be a concern for older people because hydrochloric acid production in the stomach usually decreases with age (after about age 40). If stomach acid production is inhibited by food intolerances, taking enzymes may reduce these intolerances, and help keep stomach acid production normal.

There are different types of parasites. If you suspect parasites, see a qualified doctor or health practitioner to find an effective treatment for your situation. Also, if one person in the family has parasites, it may be wise to have others close by tested as well.

The incredibly vicious viruses

A number of researchers are investigating the relationship between viruses, development disorders, and gastrointestinal problems. Viruses have been associated with intestinal inflammation among other symptoms (Kawashima *et al* 2000; Singh 2002; Uhlmann *et al* 2002). Where there is inflammation, there is usually pain!

And where there is repeated or prolonged pain, you risk developing a 'pain memory' or otherwise sensitizing the area.

A virus should be subdued and immobilized, lying dormant and harmless in the body. This is normally how the body builds up immunity to disease. In the gut, certain agents of the immune system in the mucosa lining usually conquer any viruses. However, if the intestinal mucosa is damaged or is deficient this might leave an opening for a virus to be re-activated, get out of control, or become industrious in the gut.

It may seem that a successful virus would soon overpower and conquer its host. Such a virus may appear to win out. But if it ultimately kills its host, the virus is wiped out too. A successful virus will milk whatever resources it needs from its host, but not to the point that the host finally dies, or cannot provide the nourishment the virus needs to keep on going. At least some individuals appear to be hosting such viruses inside them and living in this state.

If a virus is lingering, what happens now? Possibly the viruses lead to some gastrointestinal and/or neurological problems (Uhlmann 2000). There is evidence that viruses can cause dysfunction in the brain and damage the protective coating, called myelin, around the nerves. This leaves the nerves exposed and susceptible to damage (Singh Jan 1998, Oct 1998). The immune system is working at a higher level constantly. It is overburdened on a daily basis, yet cannot completely destroy or subdue the virus.

Test results for some children with autism conditions show altered immune system function (Jyonouchi et al 2001; Gupta et al 1998; van Gent et al 1997). And some children do respond favorably to antiviral medications. Perhaps viruses are existing quite comfortably in their young hosts. Dr Singh presents a discussion on why autism may have an autoimmune basis and why this most likely involves a virus, at least in some cases, in 'Autism, Autoimmunity and Immuno-therapy.' Viruses are suspect as agents in many autoimmune diseases. Identified viral possibilities include the stealth virus, herpes virus, measles, or viral encephalitis, which can produce autism-like symptoms. Remember, the pervasive developmental disorders are only diagnosed by observable behaviors and not on any specific physiological testing. There are a multitude of biochemical situations that could lead to these expressed behaviors.

So what do you do about a virus once it becomes a problem? A basic therapy against such viruses needs to focus on the immune system: improving its ability to function, strengthening it, and enabling it to work at a more typical rate and manner. Some people are seeing improvements with particular antiviral medications. However, because of the nature of viruses, this may be more of a cross-your-fingers-and-hope-for-the-best therapy. Researchers are working to improve this as best they can.

Enzymes? Our industrious pal the digestive enzyme has a little trick of its own. Enzymes, particularly the proteases, turn out to be an excellent therapy to use against a virus, working on several levels. Many viruses are surrounded by a protective protein film, something a protease enzyme can digest away. Eliminating this coating leaves the viruses unprotected and vulnerable to antivirals and destruction.

Is there any evidence that enzymes are effective in the treatment of viruses? In 1995, Dr Billigmann published the results of a study with enzyme therapy as an alternative in the treatment of the virus *Herpes zoster*. In a controlled study with 192 patients, one of the objectives was to confirm that enzyme therapy had been effective with this virus in a previous study. The other objective was to compare the effectiveness of enzymes with that of a standard drug called acyclovir. The high costs of treatment with this drug and others often meant that *Herpes zoster* patients would not receive medicinal therapy. They concluded that overall the enzyme preparation showed identical efficacy with the drug acyclovir, and thus also confirming the results of the prior study.

The *Herpes zoster* virus has been successfully dealt with since 1968 with enzymes. Enzymes are considered one of the best therapies with very few side-effects while also providing significant pain relief for the patient (Bartsch 1974; Scheef 1987). Bartsch eventually felt it was unethical to treat patients with viral conditions with anything other than enzyme therapy because the enzymes proved far superior as a treatment.

In addition, we have the research and successful experience gained in the field of treating HIV, another viral-based condition. The Medical Enzyme Research Institute has published the following conclusions regarding the results of controlled studies underway from 1985 to 1994. First, they found that enzyme therapy significantly limits the progression of the early stages of HIV disease and the patients' symptoms improve appreciably. Next, with people who are HIV positive, enzyme treatment can delay the onset of disease symptoms and, in some cases this delay continues indefinitely. Then, the appearance of infectious disease, and possible malignant disease, has become less common than in the control groups. Lopez (1994) gives a complete description of how enzymes assist with HIV/AIDS conditions, how they interact with the immune complexes and other factors in the disease, with scientific references.

Several other studies showed that enzyme therapy resulted in less symptoms, slower progression of the HIV viral infection, and greater improvement in the patient's condition than either controls or groups receiving other standard drug therapy, such as AZT, at times by a significant margin (Jager 1987, 1990; Kaye 1989; Stauder 1988, 1989, 1990; Ransberger 1988). It seems enzymes are a natural fit as part of any program aimed at treating viral infections.

Enzyme therapy is a proven solid and safe therapy found very effective with many pathogens, whether they are viruses, yeast, bacteria, or parasites. Fortunately, enzymes are very compatible with other therapies and not nearly as expensive. Then, once you weed out the bad stuff, you need to add back the beneficial micro-organisms.

The probiotic short course — The good guys

Probiotics are an essential key to maintaining good digestive health. Providing a steady stream of probiotics — the beneficial microbes we need — is a wonderfully simple way to keep intestinal bacteria, yeast, and other inhabitants in proper balance. The major benefits of adding probiotic organisms to the diet are:

- boost to the immune system
- vitamin synthesis
- improved digestion
- increased nutrient absorption
- improved resistance to allergies/intolerances
- production of digestive enzymes
- regular elimination
- detoxification and protection from toxins
- cancer-protective effects

- reduced risk of bowel disorders and diseases
- inhibition of disease-causing organisms/reduction of pathogens
- maintainance of the proper pH

Probiotics provide other benefits, such as producing highly favorable natural chemicals through fermentation. Scientific studies over the last 50 years show that probiotic organisms can improve the nutritional quality of foods, produce natural antibiotics, anticarcinogens, and substances that break down and recycle toxins for their human host (Linskens *et al* 2001). R. D. Rolfe summarized and reviewed the role of probiotics in the beneficial control of gastrointestinal health from 85 different scientific studies (Rolfe 2000).

That is an incredibly solid list of benefits. Go back right now and read over the list again v-e-r-y s-l-o-w-l-y. Think of these items in terms of maintaining gastrointestinal health and how they relate to the benefits of enzymes. I'll wait.

Probiotics come in a smorgasbord of types and strains. You really only need a couple of key good ones for core intestinal health. When deciding on which probiotics to use, choose an extensively researched product or strain. Each type of probiotic contributes a little differently.

The strains of probiotics known as *Lactobacillus acidophilus* is one of the most important members of the gastro-intestinal tract. Scientific studies verify this strain is immensely effective in maintaining good gut health. *Lactobacillus acidophilus* bacteria reside mostly in the small intestine. *Bifidobacterium bifidum* is an example of a strain that inhabits the large intestine (colon). Many strains of *Lactobacillus* are available as well as other probiotics.

There is one strain of *Lactobacillus* called Lactobacillus GG. It is sold in the United States as Culturelle. Lactobacillus GG is an extensively and exceptionally well-researched strain. It shows excellent results with improving the environment in the intestines. Over 141 research studies on this strain are available.

Mixed opinions prevail on whether taking a mixed-species product or a single species is better. It probably depends on the nature of the problem. You may decide to choose a single strain of friendly flora because of the proven effectiveness of that particular probiotic strain. An example would be taking *Lactobacillus* for the small intestine, or

Lactobacillus bulgaricus and *Streptococcus thermophilus* to assist with colon problems. Taking a strain for its specific properties can be very helpful. However, over an extended period of time, you may want to include other strains just for good measure.

Generally, you need higher, more therapeutic doses of probiotics when first addressing gastrointestinal symptoms. Probiotic strength is measured in CFUs (colony forming units) per capsule. For therapeutic benefits, references vary widely from 250 million to 20 billion viable organisms per day. This may depend on the particular strain, quality, and product brand. Some children with severe gastrointestinal disorders have seen improvement taking from 30 to 60 billion CFUs daily. Check with your health practitioner, or you can start slowly and build up to a level you feel is most beneficial.

Look for the manufacturing date stamped right on the container. Probiotic products may lose a lot of potency after four to ten months so date is important. Most products last longer when refrigerated, although some products do not require refrigeration. Talk to others, especially other parents and adults, about quality issues, and what works best in particular situations. Capsules are the preferred form for probiotics instead of tablets or loose powder. Capsules provide more protection from contamination, oxygen, and moisture helping maintain organism integrity.

When to take probiotics also varies by brand. Always check the label and note the recommendation of the manufacturer for optimum performance from that product. Some labels say to take on an empty stomach, while some say to take with food so the food can buffer the organisms. Some say to take in the morning because of stomach acid content, whereas you can take others anytime. The acid and salts in the gut may harm certain probiotics. Manufacturers take this into account when designing a formulation and preparing the capsules. Some capsules are specially coated so the microorganisms will safely reach their destination, which may be at different spots in the digestive tract.

Some people just cannot tolerate any probiotic supplements at all. Or find none that benefit them. A whole food source, such as properly prepared yogurt and other fermented foods with live cultures, is a good alternative. Whole foods provide extra benefits that supplements do not.

A key benefit of probiotics is the fact that certain probiotics secrete digestive enzymes, particularly proteases, lipases, and lactase. Lactase is the enzyme needed to break down the milk sugar lactose. Some of the benefits of the probiotics are probably due to the improvements brought about by this increase in digestive enzymes in the gut. Enzymes and probiotics work very well together. Enzymes clean out the harmful things in the gut and the probiotics repopulate the good microbes.

Enzymes and probiotics

At this time, the literature is inconclusive on whether to give probiotics and enzymes together or separately. In the enzyme group, the general recommendation (at this writing) is to give enzymes and probiotics at separate times – enzymes at the beginning of the meal and probiotics at the end of the meal or any other time that is at least an hour apart from the enzymes. Although most enzymes do not adversely affect most probiotics, a few parents commented that they saw improvement by giving these two supplements separately. Some product manufacturers also recommend this. It most likely depends on the products, formulation, and what is going on in a particular person's gut.

Some enzyme formulations and some probiotic formulations may interact more than others. Here are two possibilities.

1. Probiotics consist of a lot of protein, and proteases may interact to some degree, or cause some breakdown.

2. Some enzymes may prevent the probiotics from achieving the best attachment in the gut (and thereby inhibit optimum colonization).

On the other hand, probiotics produce enzymes including proteases and live quite well in the digestive tract naturally. So it is easy to speculate that enzymes will not adversely affect these probiotics. Certain probiotics may be susceptible in the small intestine but not the colon, or in the colon and not the stomach. You can also checkwith the manufacturers of both the probiotic and the enzyme.

The take-home message is that if you give enzymes and probiotics separately and see better results with that timing, go with that. If you see better results by giving them together, then stick with that.

Testing for intestinal inhabitants

You can check for what is populating the neighborhood of your intestines through a test called the Comprehensive Stool and Digestive Analysis, available through most commercial laboratories. These tests can be a good guide, but are not 100 percent conclusive. You may benefit from a good probiotic, antibiotic, or yeast treatment even if the tests suggest you do not need one. It may also tell you about stomach acid production and pancreatic enzyme insufficiency.

Treating encopresis

After many trips to the pediatrician and other specialists, I located the term 'encopresis' and how to treat it while searching on the Internet. Although I had never heard of it before, it is more common than you might guess, especially in children. It just is not the type of topic you routinely bring up over dinner or put in a family holiday newsletter. Encopresis is something to be aware of when there are possible intestinal health problems.

Encopresis can be thought of as constipation gone out of control. When the waste in the colon builds up for any number of reasons, you have constipation. If this is not relieved reasonably soon, the person can becomed impacted. An X-ray can show if there is impaction, if you want one done. At times, the solid waste builds up and causes the colon to expand to accommodate the growing mass. The colon can expand up to four times its typical size. Liquid waste may seep around the solid mass as well. This leads to streaking or staining in the underwear, smaller 'droppings' coming through, or what appears to be alternating constipation and diarrhea. If you see any of these, consider encopresis and deal with it immediately because toxins are stagnating in the colon.

In our case with Younger Son, the polluted environment in the colon appeared to be killing off the beneficial bacteria and allowing an overgrowth of the harmful bacteria. This produced the incredibly strong smell. You may also notice strong body odor or bad breath even if the person just took a shower or recently brushed their teeth.

To treat encopresis, you have some options. If there is serious impaction, consult your medical care professional for ideas. The usual treatment suggests giving a laxative or enema to start breaking up the mass and eliminating it. I favor the laxative by far personally. I chose to go the very natural route and use magnesium, which works as a laxative at higher doses. Because magnesium deficiency is so common, I thought this supplement would help with any additional magnesium my son might need at the same time. Some people opt to give one of the laxative products from the pharmacy or grocery store and use according to directions.

Next, plan on keeping a very definite schedule of when the person goes to the bathroom to 'try' and see if anything comes out. The person should go to the bathroom about ten minutes after eating breakfast and dinner at least, and any other time after eating you can remember or it is convenient. The person should 'try' for about ten minutes long. This part is very, very essential to the program because it takes advantage of the body's natural digestive movements, and helps the body to get back into a proper elimination pattern.

My son insisted he could not feel when he needed to go to the bathroom. At first, I thought this was very strange. As it turns out, this is very common and the person genuinely cannot feel the sensations. This may be due to the person having sensory integration dysfunction, as I suspect my son had because he was so insensitive to other things. When he initially did not have the sensory feedback of when he needed to go, the constipation built up, and then the problem perpetuated itself. Another route is that you may become constipated first, and as the colon stretches out, the nerves are no longer close to and in rhythm with the amount of waste. Even if the mass is eliminated, the colon is so stretched out that it will not sense the build up of more waste until another massive amount accumulates. The problem again perpetuates itself. Scheduling times to 'try' is very important, whether the person feels like he needs to go or not.

Adjust the diet to favor looser stools. Include more fiber and pure water (here's that pure water again). Many people also include something like mineral oil, commercial fiber products, or magnesium on a daily basis to keep things moving. I like magnesium to help with any deficiencies at the same time. This has worked quite well for us and I do not have to worry about giving too much. I also chose to give extra Peptizyde between meals to help clear out any rubbish, gunk, and bacteria that may have built up. I could tell this was helpful because as soon as I stopped the enzymes the problem became worse.

This anti-encopresis program needs to be kept up for a good six months or more. It is very common to see good results by the third or fourth month, and then give it up because the person appears to be doing well. What happens is that you may have reestablished a good elimination pattern, but the colon has not yet contracted back to its original size. Some people are also very prone to becoming constipated and the situation can recur easily. So it is good to plan on six months or more.

In our case, Younger Son regained his feeling in his gut about the third month, and he was doing quite well at the fourth month. This corrected the encopresis, but I still needed to treat the overgrowth of unwelcome adverse bacteria that had established themselves. Culturelle probiotics are excellent at handling this bacteria. I gave three capsules per day until we used the antibiotic to wipe the bacteria out. Then we went down to the typical dose of just one capsule of the 20 billion colony forming units strength per day. Giving the enzymes in between meals was not necessary after the antibiotic either.

If the person has sensory integration issues, there are exercises available that can help regain better feeling in this area. Sensory therapy may be a good option to explore in case the colon problems leave a distored sensory or pain memory.

The five-month mark came. No crashing.

CHAPTER 15

Why You Should Eat
Like A Pig!

Enzymes are a very common and long-standing addition to the diets of livestock animals and domestic pets. As in many areas of mammal health and nutrition, the knowledge, use, and refinement of digestive enzymes in animal science is far ahead of that with humans.

Some animals are ruminants, such as cows, sheep, and llamas. They have more than one stomach. Digestion is a bit different in animals with a single stomach, such as pigs, poultry, cats, and rabbits. Animals with one stomach cannot digest compounds called non-starch polysaccharides, commonly known as 'roughage.' Ruminants can digest this insoluble fiber in their additional stomachs. The animal with a digestive system most like humans are pigs, so this discussion will focus on non-ruminants.

A main reason so much attention is given to maximizing digestion and nutritional efficiency in livestock is simply the bottom line. A producer pays for feed and wants to get the best value out of it. Anything passing through the animal unutilized is wasted money, time, and effort. In addition, it costs cash to shovel, haul, and dispose manure. Poor digestion is, literally, money going down the drain.

Think of livestock production as a well-oiled biological machine. More food in, quick and optimal digestion of feed, maximize nutrient absorption, convert energy in feed to energy the animal can use, minimize waste – and keep it *moving* for the best rate of growth. This strategy results in the most efficient and cost-effective method to grow the animal. Anything compromising the animal's health is pounced on immediately and remedied. Human health care may not have the exact same motivation or goals as agricultural producers, but animal science has produced a wealth of knowledge on digestion and nutrition we can benefit strongly from and apply.

Enzymes, animal science, and veterinary medicine

Farmers are keenly aware that excessive material in the colon is food for harmful bacteria. They also know these pathogens can severely compromise an animal's overall health. Excess ammonia given off by adverse bacteria is considered very detrimental not only to the animal but to the people working closely with them. Digestive enzymes are frequently used with animals to cut down on ammonia. The idea is to give enzymes that can break down the hard-to-digest plant fiber and roughage so the nutrients and energy benefit the animal and leave little left over to feed adverse bacteria. Commonly used enzymes include cellulases, hemi-cellulases, and xylanases.

With more undigested fiber in the intestines, the rate at which the food passes through is prolonged (the transit time is too long). Why avoid food lingering in the gut too long? Why 'keep it moving'?

For starters, a thick, sticky, dense mass of digestive contents moving slowly thorough the intestines can provide a stable environment for microbial growth and proliferation of bacteria. High intestinal bacteria populations irritate the gut lining, damaging microvilli, and decreasing nutrient absorption. The slower rate of food passage provides an almost protective barrier rich in nutrients which helps bacteria grow (McDonald *et al* 2001).

An example study is from Sarosiek, Slomiany, and Slomiany (1988) who monitored pig intestinal linings for feed transit time and consistency. *Campylobacter pylori* became established and weakened the gastric mucosal barrier. This bacteria caused a rapid degradation of mucus proteins, which at the end of 48 hours decreased the mucosal lining by 36% and increased gut permeability.

Hopwood, Pethick, and Hampson (2002) found that increased density, stickiness, and slowing of the digestive matter in the intestines stimulated the proliferation of the toxic pathogens *Escherichia coli* and *Brachyspira pilosicoli* in weaned piglets. Sluggish digestion:

- Results in less mixing in the gut; gut less able to stir contents
- Prevents enzymes from reaching food parts for breakdown
- Decreases contact of nutrients with gut wall for absorption
- Decreases absorption of nutrients (poor growth)
- Increases undigested feed and water exiting the animal
- Decreases feed conversion into nutrition and energy
- Limits feed intake; reduces feeding (growth slows)

Pig studies show intestinal bacteria counts are higher with wheat, barley, and rye diets than with the more highly digestible corn-based diets. So, the most attention goes to increasing the digestibility of various carbohydrate fractions of cereals and plant proteins. Enzymes are usually added to the diet to aid the digestion of wheat, barley, and rye. (Hopwood *et al* 2001)

The effectiveness of such supplemental enzymes is verified by Pan *et al* (2002). This team examined growth performance, nutrient digestibility, and intestinal bacterial populations in piglets. In summary, all factors improved significantly with the added enzymes, along with the bacterial populations being modified for the better.

Giving digestive enzymes supplements the animal's own internally produced enzymes. Consuming more feed may overwhelm the animal's ability to produce sufficient enzymes either from its own tissue or through the micro-flora in the gut (remember that probiotics also produce digestive enzymes). Broiler chickens have a higher rate of feed intake than other poultry, and studies show they respond well to additional enzymes (Danicke *et al* 1997; Oderkirk 1996).

A study by Gracia *et al* in 2003 took an extensive look at the influence of giving starch digesting amylase enzymes on the digestion and performance of chickens on a corn-soy meal diet. The amylase improved daily gain by 9.4%, and feed nutrient and energy absorption by 4.2%. By the end of the trial, the chickens consuming amylase ate more, grew faster, and grew healthier than the controls.

In addition, the enzyme containing diet reduced relative pancreas weight but not the weight of other organs. This is consistent with the studies of Pottinger's cats and Dr Howell's extensive research on how well supplemental enzymes relieved the workload on the pancreas. Even more interesting is that these current studies continue to verify and validate that work done so many years ago.

Enzyme supplementation may potentially improve the nutritional value of feedstuffs (Oderkirk 1996). Enzymes are given to:

• Improve nutrient absorption and utilization
• Maintain or improve performance on lower quality diets
• Decrease feed and supplement costs
• Reduce cost of treating illnesses
• Widen range of foods that can be given
• Overcome sluggish digestion and anti-nutritional factors
• Decrease wasted nutrient and water excretion

These benefits are identical to the objectives for human nutritional health care. So apparently these goals are the foundation of universal optimal digestion and nutritional wealth. And enzymes make it happen!

Carbohydrates - Something to chew on

A dietary approach that tackles the adverse bacteria problem with severe intestinal problems in humans is the Specific Carbohydrate Diet (SCD). It was developed for those suffering from serious bowel dysfunctions such as Crohn's disease, ulcerative colitis, celiac disease, and irritable bowel disease (and even some with autism). Bacteria, and perhaps yeast, are significant problems with these conditions.

The SCD is based on two related issues. First, some of the body's digestive enzymes are located at the mucosal lining. If the gut lining is injured for any reason, these enzymes may no longer be available in sufficient quantities, or at all. These enzymes are lactase (for milk sugar lactose), several other disaccharidases (sucrase, maltase, isomaltase for complex sugars) and some peptidases (break small proteins). Recall how many individuals with neurological and intestinal problems were also found to be very low in lactase and disaccharidases.

The second consideration is that carbohydrates have the greatest influence on intestinal microbes. Any carbohydrate not sufficiently digested and absorbed remains a source of food and fuel for microbes.

Adverse bacteria produce harmful acids, ammonia, and toxins that injure the gut and pollute the body. This starts the vicious cycle where further damage to the gut only promotes further lack of digestion, more microbial growth, more neuro-toxins, and so on.

The SCD specifically eliminates carbohydrates that rely on the intestinally-derived enzymes. The theory is the starches and sugars allowed are quickly absorbed leaving virtually nothing for harmful microbes to feed on and thus die. To what extent this actually occurs depends on the individual situation. Just eliminating foods does not necessarily promote gut healing or completely eliminate the pathogen problem. Some additional pro-active healing elements may be needed.

An interesting part about SCD is that it advocates a special yogurt and certain cheeses made from cow, goat, or other milks. Most people doing SCD agree the real healing begins when these dairy products are added. The dairy products are a key factor as they provide needed probiotics, enzymes, minerals, anti-microbial factors, and other elements in a whole food form. These beneficial components in dairy contribute to restoring intestinal health. The preparation process eliminates the lactose in these dairy foods.

People concerned about casein proteins being fully digested have found this may not be a problem on SCD. It might be because the probiotics used can break this casein down (research shows they can produce DPP IV activity, lactase, and other enzymes). The yogurt and cheese making processes also degrade the casein proteins.

The previously given animal studies confirm veterinary medicine has been aware of this bacteria-based vicious cycle for some time, and enzymes are a major part of the plan to counter it. Enzymes may be accomplishing the same thing as the SCD does with food eliminations, at least to some extent. Afterall, this is exactly why these enzymes are a long-standing part of animal care. For enzyme products for animals, consult your veterinarian or animal nutritionist.

The cat's meow

Pet owners take a personal interest in their animal's health. Veterinarians know that just like in humans, the natural ability of dogs and cats to make their own digestive enzymes slows down with age. Many enzyme products for pets contain the same mix of enzymes of the same quality grade as are in products for humans.

And like their human counterparts, pet food and pet diets are far more processed today than diets containing more raw food in years gone by. This leaves our furry friends in the same predicament we face – a cooked diet deficient in enzymes placing more stress on the body. Veterinarians more readily recommend high-quality enzymes for their four-legged patients. Some animal nutritionists suggest adding more raw foods into the diet as well. Maybe a little finely ground raw meat, raw vegetables, and even a bit of whole grains such as oats. Additional enzymes or raw foods are advised for cats and rabbits so the enzymes can break down furballs.

Whenever you change the diet, do it slowly so your pet won't experience diarrhea as the microbe population shifts around. Unwell pets, like unwell people with sensitive neurology and digestive tracts, respond better when their systems are rebalanced at a pace that will be helpful, not cause more harm in the process. Animal caregivers know that rushing detoxification or going 'cold turkey' on diets, or giving mega-doses of supplements (which isn't really prescribed or supported in animal medicine) can overwhelm the system. Recovery may not be any faster, or even slower, than a slow-and-gentle approach.

It is also notable that in pet care, enzymes are sought and recommended to help a dog or cat with degenerative diseases. Enzymes are given to increase T-cells (cancer fighting cells), strengthen the immune system, increase white blood cells, balance cholesterol, pancreatitis, diabetes, virus infections, chronic worms, reduce inflammation, and clear arteries of plaque. Same reasons as for people.

Young piglets start off on enzymes with their digestive health under constant care. Domestic pets usually do not receive enzymes when young, but only as the pet becomes older with faltering health. Unfortunately, it is this second scenario that is more common with humans. And Fluffy usually has the advantage of the vet advising enzymes much sooner and more often in the process than most doctors do for people. It isn't physicians are against enzymes – more a matter of they do not think of it. Or consider it a safe daily-living alternative one would do on their own not requiring a medical doctor's approval. Like eat an apple a day, add whole grains, or drink clean water.

CHAPTER 16

Magnesium and Neurology

Let me tell you about Matthew's head-banging. Matthew has been banging his head and humming in a droning rhythmic way since one year old. For the first year, he would *scream* himself to sleep and *scream* himself awake at every nap and every night. So, at first, head-banging was a nicer, quieter improvement. However, he continued to get out of his bed and bang his head on the floor ALL NIGHT LONG for seven long years. Round the clock. Thump, thump, thump. In the car, thump, thump, thump. On the sofa, thump, thump, thump. The pediatricians we saw said that 15 percent of all boys are head-bangers and he would grow out of it by five years old. We were afraid he was going to get brain damage because of the sheer intensity with which he pounded his head. The house would vibrate because of the force. At traffic lights, the car would rock.

If your little one ever asks you what goes thump in the night, tell him not to worry – it was only my little boy banging his head, with the vibrations reaching your house.

Why did he do that? It may not seem like banging your head serves any useful purpose...until you think of it in terms of pain management. It turns out Matthew had a good reason. I tried pounding my head repetitively on the floor and, by gosh, it really works!

Just a light hitting or using a soft object didn't help; I had to give my head a substantial whack to get good results. When you are in constant pain, or feel miserable all over, it just drives you insane trying to endure it moment after moment after moment. If you whack your head pretty solidly, you feel 'stunned' for just a split second when you don't feel the pain before it seeps back into your consciousness. If you whack, whack, whack continuously, you build up a series of stunned moments. Focusing on the no-pain moments can bring some much needed relief. This may not be the reason for every person who bangs his or her head, but it seems to be at least one.

Another view is that when you are in such pain or discomfort, a diversion can offer some needed relief. Whacking your head, scratching your skin, or other activities may cause a little bit of pain, bruising, or even bleeding. However, that pain is targeted and it helps one to focus on something besides how miserable you are in general. You might say it is 'calming.' So it is a way of diverting attention from one pain to another different one.

Think of being attacked by a tiger. If the tiger scratches you, gashing your leg, your primary goal is to get away from the tiger first even though you are severely bleeding and in pain. Once you are free and clear of the tiger, *then* you can focus on your injury. Someone watching from a safe distance not seeing the tiger at all, would wonder, 'Why is that guy running around with a gashed leg?! He is only inflicting more pain on himself and making the wound worse.' I suspect many people with neurological discomforts are busy fighting tigers that other people cannot see.

Matthew wouldn't or couldn't even sleep in a bed until he was five. Even after we started him on Zoloft, he would stay in the bed some of the night, then get out of bed during the night, and bang his head on the floor. Thump, thump, thump. At seven years old, we started amitryptiline for migraine head pain and with that he would sleep in the bed all night, but still bounced his head on a pillow during the night. Thump, thump, thump. About 75 percent of the head-banging during the day stopped with the amitryptiline. We did not notice any difference in the head-banging when eliminating foods.

One week after starting Peptizyde, he stopped banging his head during the day and night. He stopped. He just says, 'Goodnight, Mommy' and goes to sleep. No more nightly battles to get him in bed – he prances right off to his room. Often he closes his eyes, bounces his head gently on the pillow a little bit, and then goes to sleep all night long staying in the bed. At other times, he may rock to music or when he is concentrating. However, he now does this appropriately, not compulsively. If I ask him to stop, he will. To tell you the truth, it was a little weird in the beginning and we needed to adjust to the pleasantness. We are very happy. So is Matthew.

Often I hear the description that people with autism spectrum conditions are 'living happily in their own little world.' I do not know about other people, but Matthew was not happy in his own world. He was miserable. He was in pain. He was coping as best he could. I understand because I suffered the same. If someone has a neurological condition and does not feel they are suffering, or are miserable, I am glad they are not hurting. However, some people are and the resulting behavior or symptoms may be displayed as huge sensory problems, behavioral disability, attention deficit, or something else. Reducing any physical distress should not be overlooked.

At the first meeting with the Incredibly-Helpful-Neurologist, she recommended magnesium (and baclofen) right away to help with the migraine pain - which then assisted with the disrupted sleep and sensory issues. Magnesium is advised for helping reduce muscle spasms, hyperactivity, and anxiety; and a known treatment for migraine. It is commonly deficient in a standard diet of processed foods. Magnesium relaxes and reduces some of the physical tension. Even now, when Matthew gets too 'jumpy' from time to time, or starts to complain of a headache and bounce his head, some extra magnesium helps.

Magnesium – The science behind it

Magnesium along with calcium and sulfur, is a mineral needed in larger quantities rather than the very trace quantities needed of most other minerals. A remarkably interesting thing I found was the clinical indications of magnesium deficiency. Magnesium deficiency is associated with: AD(H)D, Alzheimer's disease, anxiety, asthma, arthritis, autism, autoimmune disorders (all types), chronic fatigue syndrome, chronic pain, colitis, constipation, Crohn's, depression, diabetes, fibromyalgia, food allergy/intolerances, gut disorders, peptic ulcer, headaches, hyperactivity, hypertension, hypoglycemia, insomnia, irritable bowel syndrome, lupus, migraines, multiple sclerosis, muscle cramps, fatigue, Parkinson's, PMS, stress, and tension.

That familiar list again! It appears magnesium is deficient with most autoimmune, neurological, and gastrointestinal conditions, and supplementing it can help symptoms, at times improving them tremendously. Numerous studies confirm how extensive magnesium deficiency can be in children and how it contributes to neurology.

Kozielec *et al* (1994) found that the magnesium, zinc, copper, iron, and calcium levels in blood plasma, urine, and hair in 50 children aged from four to 13 years with hyperactivity were lower compared with a control group. In 1997, he found magnesium deficiency in 95 per cent of 116 children (94 boys and 20 girls), aged 9 to 12 years, with recognized ADHD in both hair and blood analysis (Kozielec and Starobrat-Hermelin 1997). Supplementing magnesium may substantially help those with hyperactivity. Fifty children diagnosed with attention deficit hyperactivity disorder and magnesium deficiency received magnesium supplements for six months. This produced a significant decrease in the hyperactivity (Starobrat-Hermelin and Kozielec 1997). A following study gave similar results when 75 children known to have ADHD and magnesium deficiency received magnesium supplements, also resulting in notable decreases in hyperactivity (Starobrat-Hermelin 1998)

In double-blind trials with 60 autism spectrum children, researchers found that improvements appeared with vitamin B6 only when magnesium was added (Martineau, Barthelemy, and Lelord 1986; Martineau *et al* 1985). Adding magnesium to the diet might be the key factor providing substantial benefits.

Symptoms of magnesium deficiency

The symptoms of magnesium deficiency are irritability, tantrums, seizures, insomnia, muscle cramps/twitching, hyperactivity, and poor digestion among others. Magnesium is essential for proper electrolyte function, over 300 enzyme functions, and calcium absorption. Calcium and magnesium need to be in the proper balance. Many people may be calcium deficient not because they do not have enough calcium, but because they do not have enough of the needed magnesium to assist in calcium absorption.

Magnesium is vital for the many enzyme systems that help restore normal energy levels as well as supporting nerve and muscle functions. These roles lead magnesium to be helpful in many cases of nervousness, anxiety, insomnia, depression, and muscle cramps. Magnesium is a common part of treating autism or hyperactivity.

I also found it very interesting to read that one of the primary sources of dietary magnesium is whole grains and cereals. If one goes 100 percent gluten-free, you lose a primary source of magnesium, and could become *very* deficient, especially if also supplementing with extra calcium to make up for being casein-free. Higher amounts of magnesium may cause a laxative effect that may or may not be desirable (the basis of consuming milk of magnesia or Epsom salts).

Forms of magnesium

Magnesium supplements fall into two basic groups:

1. soluble forms or organic (aspartate, malate, glycinate, citrate, succinate, etc.)

2. insoluble forms or inorganic salts (chloride, carbonate, oxide)

Taking magnesium between meals on an empty stomach facilitates absorption. The soluble forms are fairly equally absorbed and as a group are much better absorbed than the insoluble group. You may occasionally find something with ionic magnesium and this is also highly absorbable. The insoluble group is far more likely to cause loose stools than the soluble group. If magnesium nutrition is what you want, go for the soluble group.

Many products are a mixture of magnesium forms. Some are a specific type of magnesium, such as magnesium citrate, but products marked just 'Magnesium' were usually a mixture. Some products contained as many as five different forms of magnesium. Personal favorites for headaches and calming are magnesium citrate or malate, which seem to work quickly. Other soluble forms may give very good relief too. We use homemade magnesium sulfate cream (Epsom salts) as it is soluble, supplies both magnesium and sulfur, and is inexpensive.

Recommended dosing

Many sources saying the usual recommended daily amounts of magnesium for healthy people are actually too low. Most nutritional authorities suggest a ratio of twice as much calcium as magnesium for regular function. To correct a deficiency, a one-to-one equal match may be more helpful. A few general reference books on supplementing recommended 1000 milligrams of magnesium per day for attention deficit conditions, anxiety, migraines, sleep problems, autism and some autoimmune dysfunctions. These are probably for an adult, so a child should have about one-third or one-half the amount. This puts the dose around 300 to 500 milligrams of elemental magnesium per day. Giving supplements in smaller more frequent doses can be more effective and better absorbed than in one larger dose once a day.

The general upper limit of magnesium is when stools get too loose, and I read several places that no level of toxicity is known. The body does not store magnesium as it does other nutrients like calcium. It excretes what it does not use. Minerals need to be in balance because they tend to compete with one another, particularly magnesium, calcium, and zinc.

A few comments on calcium

When considering calcium, first see if a magnesium deficiency needs correcting. You may be getting enough calcium already but the body is not able to use it because it lacks the magnesium needed for calcium absorption. If you are low in magnesium and supplement with even more calcium, the calcium deficiency problem only becomes worse.

Next, look for a mineral source the body can easily absorb. Just as with magnesium, look for water-soluble, chelated, or ionic forms (such as citrate, malate, lactate, etc.), not the carbonate or oxide forms. This tends to be especially important for those who have digestive and absorption dysfunctions anyway. The least expensive, good quality source I found was NOW brand calcium citrate at the local store.

Calcium is best taken between meals, also. Although many sources recommend taking a carbonate form with meals, this can slow or impede digestion (this is why most over-the-counter antacids consist of calcium carbonate). Calcium can be calming so taking this at night may promote better sleep. Giving doses of 500 mg calcium or less at a time is advisable as the body doesn't absorb more than that at once.

Other nutrients are also needed for good calcium uptake, such as vitamin D and K and potassium, but these usually do not need to be supplemented in addition to a regular diet. Since there is no total consensus on any part of calcium supplementation, the consumer needs to be very informed in order to find a product that meets their needs, and educated in knowing how to interpret the mountain of literature available.

Reading the label – How much mineral are you actually getting

You need to balance out how absorbable a product is versus the quantity of the mineral you are getting versus the price. The Food and Drug Administration (FDA) says the label should list the amount of the *element* by weight. Let's use calcium as an example. The label lists the form that the calcium comes in inside parentheses beside the weight. So, if the label says 'Calcium 600 mg (as calcium malate),' then you know that in each serving you are getting 600 mg of elemental calcium from the source compound calcium malate. The weight of the malate compound would be much more. You are only interested in the weight of the elemental calcium because that is what your body needs and uses.

If the label does not use this format (then they are not in compliance with FDA guidelines for one thing) and just uses the form of calcium and then the weight, such as calcium carbonate 1000 mg, then the weight listed is the weight of the entire compound and not the weight of just elemental calcium. In this example, you would need to multiply the weight of the compound by the percent of calcium in the compound. Calcium carbonate contains 40 percent calcium by weight. So, 1000 mg x 40 percent = 400 mg of elemental calcium. That is the amount of elemental calcium mineral you would be getting from 1000 mg of calcium carbonate.

If one product says each serving contains calcium 250 mg (calcium carbonate), and another bottle says each serving contains calcium 250 mg (calcium citrate), you are getting 250 mg of elemental calcium from either bottle per serving given. At this point, look at the form of the calcium and whether it is tablet, capsule, powder, or liquid, and how much constitutes a serving if this makes a difference to you.

Look for supplements that say 'purified' or have the USP (US Pharmacopoeia) symbol. The USP symbol, which is voluntary among vitamin and minerals manufacturers, means that the supplement meets certain criteria for quality, strength, and purity.

Solubility, absorption, and bioavailability

Solubility means the ability of a product to go into solution. The body can only use elemental calcium or magnesium and so the compound must break up into the individual parts. A product must dissolve first before it can be absorbed.

A supplement should disintegrate in your stomach in an adequate amount of time for it to be useful to your body. To see if your mineral supplement disintegrates easily, place it in a bowl or glass in about 6 oz of vinegar or warm water. Stir occasionally for 30 minutes. It should be completely dissolved. If not, it is likely that this is how it will remain in your body and you should look for something different. Most water-soluble forms dissolve readily in the gut. Check tablet forms because the binders holding the tablet together may inhibit this reaction. Powdered products usually dissolve quickly, and the liquid supplements already have the compound in solution.

Absorption means the ability of the elemental mineral in the product taken into the body in an acceptable form. Many manufacturers claim their minerals are easily absorbed. There can be two different interpretations of this:

- Absorbed into the system – Anything that is one micron or smaller will pass through the stomach wall and go into the bloodstream. One could say the product has been 'absorbed.' Most absorption occurs in the small intestine.

• Absorbed into the cells – Just because something passes through the stomach wall into the blood does not necessarily mean it is usable. To be used by the body, the minerals must be able to enter individual cells. The mineral needs to be angstrom size, one million times smaller than a micron. Ionic forms are readily available. Other forms may or may not reach this size.

Bioavailability refers to how well the body uses the elemental mineral in a desirable way, not just taken into the bloodstream and floating around. Even if a particle is small enough to be absorbed into the bloodstream, it may not be small enough, or in the right form to be used at the cell level; therfore it cannot meet the body's requirement. Minerals as whole food sources may be far more bioavailable than in synthetic or isolated forms.

Some mineral forms will simply move through the digestive tract and exit the body. Nothing is gained and nothing is lost except the cost of purchase. At other times the consequences are more serious. Certain minerals, like calcium, tend to build up in the body to toxic levels. That leads to calcium deposits which is associated with kidney stones, types of arthritis, arterial plaque, and other disease conditions. *Lynn?* This is why it is important to determine and treat a magnesium deficiency, and not just keep throwing more calcium at a perceived lack of calcium. Magnesium can dissolve any excess calcium from the body while helping any needed calcium be assimilated.

Chelated minerals During the chelation process an enzyme, protein, amino acid, or something gets wrapped around a mineral or something else. When a cell is in need of one of those nutrients, the mineral wrapped in that nutrient is taken into the cell. The outer coating gets digested and guess what! There is a mineral hidden in the center! The body did not directly identify a mineral, but our goal was achieved; the mineral was taken into the cell. Chelation sometimes increases the bioavailability to 30 or 40 percent.

The Chewies

Matthew and I both seemed to have this overwhelming need to chew constantly. Probably to sooth the irritation in our heads. Maybe as way to relieve tension from excess sensory input. I thought my son was the only kid to chew his shirts to shreds, coming home with the collars or cuffs soaked from mouthing. Or gnaw the furniture. Maybe it was a break from head-banging. I guess I wasn't much better by constantly crunching ice, hard crackers, or something similar. Or sipping on a very hot, a very cold, or a very carbonated drink all the time. I can always tell when we are not feeling well because the first symptom is to chew, chew, chew.

The Chewies turn out to be commonly associated with sensory integration dysfunction and certain mineral deficiencies. Notice that all the things we did or foods we chose directly affected the head/mouth/gut sensory system. Sensory therapy exercises for the mouth can help greatly. Low magnesium and zinc are also suggested as possible causes of the chewies, and supplementing these minerals may help relieve the chomping. The combination of medicines, enzymes, sensory therapy, and a little extra magnesium and zinc keep us from being chewed out of house and home.

Marketing of minerals

Selling supplements has become a large, thriving business. Supplements are a largely unregulated industry, which can yield very profitable margins. Larger companies selling products more expensively very often does not mean a better product, or a better value (although sometimes it might). Your local store may sell something equally good and effective at a fraction of the price. Caution and a conservative approach are wise.

This link goes to an article on the marketing of minerals. It contains a nice description of the tactics and considerations involved in selling mineral supplements. Generally, a company will sell on price, source of mineral form, quality, performance, or service – trying to convince you that any of these particular things is the most important one of all. www.nutraceuticalsworld.com/janfeb001.htm

The six-month mark came. No crashing.

Chapter 17

Sulfur, Epsom Salts, and Phenols

Dana's son was very sensitive to phenolic foods among many others (this is the same Dana that found her children were most likely celiac).

My son at age three was very hyperactive and had a lot of difficulty sleeping. I never knew how long he would be awake before he fell asleep, whether or not he would wake up in the night laughing for several hours, or how early in the morning he would be awake and ready for his day. After much research, I discovered information on phenols. Phenols are chemicals found in basically all foods. For some children, their bodies have difficulty processing the phenols into useful or at least non-harmful substances.

My son would be hyperactive and laughing, at all hours of the day or night, and have dark circles under his eyes. Some typical symptoms experienced by children with phenol processing problems include dark circles under the eyes, red face/ears, diarrhea, hyperactivity, aggression, headache, head-banging or other self-injury, inappropriate laughter, difficulty falling asleep at night, and night waking for several hours. (These symptoms are very common of yeast overgrowth as well).

So, I needed to reduce my son's phenol intake, or help his body process them. Otherwise, they built up to levels that affected his behavior and physical condition. I began by removing all high-phenol foods from his daily diet: food dyes, tomatoes, apples, peanuts, bananas, oranges, cocoa, red grapes, colored fruits, and milk. Once I removed all these foods, my son finally slept all night at least nine and a half hours each night. Because I have four children, this was WONDERFUL!

I tried adding back these foods one at a time, and I discovered that even a very small amount of any of them would send my son into a hyperactive sleepless frenzy. Fortunately, most children do not react so strongly as did my son; another of my children who also is delayed does not have phenol issues at all. Unfortunately, several of the common supplements generally recommended for autism spectrum children also contribute to phenol or phenol-type issues, including some B vitamins, DMG/TMG, glutamine, grapefruit seed extract, and several others.

After being on the casein-free, gluten-free, other food-free diet for about 18 months, I began giving my children Peptizyde and Zyme Prime enzymes. I discovered they did help somewhat with some of my son's phenol issues, for certain foods but not others. We tested the new No-Fenol enzymes for phenols, and it worked very well on these phenolic foods for my family.

Later, I began a program to remove excess environmental toxins from my children. Since then, my son now has no phenol issues whatsoever, and I can give him basically any foods he wants in any quantities without difficulties. This is very nice!

Highly phenolic foods and chemicals are problematic for many adults and children with an assortment of neurological conditions. There is some interesting research on why and what to do about it. When I participated in the groups focusing on autism spectrum, most of the conversation was about limiting highly phenolic foods, using Epsom salts, and leaving out a chemical group here or there. In the groups dealing with attention deficit conditions, the discussions centered around eliminating salicylates, artificial additives, and chemicals. Both groups seemed to be dealing with the same issue, yet the jargon was different. Surprising to me was that the attention deficit groups did not seem to be very familiar with the benefits of Epsom salts or magnesium; and the autism groups rarely discussed the Feingold or Failsafe programs – programs used very successfully for assorted behavior and physical problems, and multiple chemical sensitivities.

Sulfation – The science behind it

Dr Rosemary Waring found that most people with autism conditions have a deficiency in a key detoxification pathway. The pathway involves using sulfur in the form of sulfate (known as sulfation). The enzyme involved is phenol sulfur-transferase (PST), but the problem is thought to hinge on an inadequate supply of usable sulfate ions, not the metabolic enzyme.

Dr Waring found that most children on the autism spectrum are very low in sulfate and may be as low as 15 percent of the amount in neurologically typical people (O'Reilly and Waring 1993; Waring and Ngong 1993). According to McFadden (1996), the ability to metabolize sulfur compounds in the human population occurs in different levels, and 2.5 percent are actually 'non-metabolizers.' People with low or no ability to convert compounds to sulfate have problems handling environmental chemicals, some medications, and even some chemicals produced within the body. They include people with conditions such as Alzheimer's disease, Parkinson's disease, and rheumatoid arthritis, as well as chemical sensitivities and diet-responsive autism (Alberti et al 1999).

The PST sulfation pathway is necessary for the breakdown and removal of certain toxins in the body. This includes the processing of a type of chemical called a phenol. Phenols are a regular and necessary part of life. All foods contain some phenolic compounds. Some foods have a much higher content than others do. If the sulfation pathway is not functioning well, the phenolic compounds may not be processed out as fast as they consume them. There is a cumulative effect. When phenols, or other compounds using this pathway, start backing up in the system, the result can be a myriad of negative reactions.

Symptoms of phenol intolerance include the previously mentioned ones as well as night sweats, irritability, and skin conditions. The symptoms of phenol intolerance and yeast may be very similar because they both involve the body trying to deal with toxins.

This detoxification pathway processes phenolic compounds such as salicylates (a type of phenol), artificial food colorings, artificial flavorings, and some preservatives. To make matters worse, besides tying up the PST pathway, salicylates further suppress the activity of any PST enzyme present in the gut (Harris 1998). Food dyes also have been shown to inhibit the PST enzyme. Some of the food additives avoided by the Feingold Program can suppress certain enzymes and pancreatic secretions as well (Bamforth 1993; Li 1995).

You can unclog this 'bottleneck' in one of two ways. One is by reducing the amount of phenols and toxins entering the body. This is the basis of the Feingold Program or diet. The Feingold Program removes the hard-to-process artificial colorings, flavorings, and three preservatives (BHA, BHT, TBHQ). It also removes the most problematic of the salicylate foods at the beginning of the program, such as apples, oranges, grapes, a few vegetables and other foods, plus aspirin. Later, you may be able to add the salicylate foods back after testing them for tolerance one at a time. Eliminating these chemicals has been effectively helping many children with all sorts of physiological and behavior problems for many years, although the reasons why are just now beginning to be understood.

There is an abundance of studies in the references that specifically show that eliminating these types of chemicals significantly improves neurological problems in sensitive children and adults. The Failsafe diet is similar to the Feingold Program only it is stricter by eliminating more chemicals. Between 1980 and 1994, a number of studies using Feingold-type diets reported improvements in children with ADHD. In nine studies, the percentage of children showing significant improvements ranged from 76 to 85 percent (e.g. Egger *et al* 1992; Swanson and Kinsbourne 1980). One study was outside this range, with a reported 50 percent of children showing significant improvement (Kaplan *et al* 1989). Those are substantial numbers.

A literature review by Kidd (2000) concludes that although the exact cause of attention deficit conditions is unknown, the current consensus is that genetics plays a role. Other major contributors include adverse responses to food additives, food intolerances, sensitivities to environmental chemicals, nutrient deficiencies, and exposures to neuro-developmental toxins.

This sounds exactly like the factors contributing to autism, migraines, and sensory integration issues. I was particularly interested in a study by Alam *et al* in 1997, which also involved Dr Waring.

They looked at the activity of PST with regards to migraine and tension headache and concluded that PST activity may be a factor in the cause of migraines. Another connection. Apparently, there is overlap at the biological level.

The second method of enhancing the detoxification process is to supply more sulfate. This increases the amount of toxins processed out. Sulfate ions may not be absorbed well from the gut, so simply giving more sulfur directly by swallowing supplements may not produce satisfactory results. Some people have seen improvements by supplementing with the sulfur-containing amino acids cysteine and taurine, or MSM (methysulfonylmethane), or by using one of the many commercially available MSM creams. However, others have not found this tolerable. This may be because their body is unable to convert the sulfur to the needed sulfate form. Most people do see improvement with Epsom salts because the form of sulfur in the Epsom salts is already sulfate and readily available to the body.

More reading and resources on sulfation, phenol intolerance, and additives is at: *www.enzymestuff.com*. More information about the Feingold Program is at *www.feingold.org*. This site also contains information and research on the possible behavioral and physical symptoms from various food and environmental chemicals.

What are Epsom salts and how do they work?

Epsom salts are magnesium sulfate. Salts are just molecules that form because the parts have opposite electrical charges that bind together. Magnesium has a positive charge. Sulfate has a negative charge, and performs all sorts of unique biological functions. The two elements dissociate in solution (English translation: break apart and separate in liquid). Epsom salts are available very inexpensively at most local grocers or health food stores, or in bulk at agricultural supply stores.

The magnesium and sulfate in the salts are absorbed into the body through the skin. Because the sulfur is already in the sulfate form, it does not need to be converted like other forms of sulfur do. Sulfate is thought to circulate in the body up to about nine hours. Any Epsom salts left on the skin may continue to be absorbed as long as it is still on the skin, offering continuous 'timed-released' input into the bloodstream – like medications given through skin patches. A large number of people on a typical 'modern' processed diet are very deficient in magnesium, which Epsom salts also supply in a highly available form. Main effects of insufficient magnesium are hyperness, irritability, anxiety, poor sleep, and muscle spasms. So the salts may provide two-way assistance.

A historical note: people traditionally sought out mineral springs and natural sulfur waters as health spas and rejuvenation therapy for various illnesses and healing. Natural sulfate and magnesium sources.

How to give Epsom salts

Here are several methods for giving Epsom salts. The ratio is not exact, just what seems to get the salts dissolved and on the skin.

Epsom salt baths – Most people use about one to two cups per tub. Dissolve the salts in hot water first and then fill the tub to about waist deep, as warm as possible. The amount of salts you may find works best will depend on the individual tolerance, the temperature of the water, and the size of the tub. The warmer the water and larger the tub, the more salts will dissolve. If you see negative reactions, such as irritability or hyperactivity, then decrease the amount of salts. You may need to start with as little as one tablespoon of salts, and work up gradually. Epsom salts baths are very calming for most people. This works well just before bedtime. Most guides say to soak for about 20 minutes or more. It is okay to let the salts dry on the skin. You may notice a dry clear-white powder. If it is too itchy or irritating, just rinse it off. If the skin feels too dry, use lotion or oils to moisturize. Loose stools may result if children drink the bath water.

Homemade lotion – This is my favorite at the moment. Cheap and easy. Heat some Epsom salts with a little water to dissolve them. I put about one teaspoon of water in three tablespoons of salts and microwave for a minute or so. Add more water if necessary. Then mix this into around four ounces of any lotion or cream you like. I have used suntan lotion, handcream, cocoa butter, body lotion, aloe vera cream, whatever I find that is on sale or inexpensive without the chemicals I am trying to avoid. Good buys are at the local grocer in the lotion section. Apply to skin anywhere as often as desired. Some new commercially prepared Epsom salt creams are available but can be very expensive and may contain chemicals that are not tolerated.

Poultice or skin patch – You can mix some Epsom salts and whatever kind of lotion the person can tolerate into a paste. Put this paste on a large bandaid and apply to the skin. The salts will soak into the skin.

Spray – Mix one part salts and one part water (add more water if the salts are not dissolved) and put in a spray-squirt bottle. Mist the person's chest and/or back and let it dry on the skin. This method works well in the summer.

Footbath – Mix one part salts to two parts water (or more so the salts dissolve) and let the person soak their feet in it. My boys would soak their feet about 30 minutes while they did reading or homework.

Sponge – A solution of one part salts to four parts water works well. Dampen a sponge in the mixture and apply to any part of the body.

Epsom salt oil – Neither of my sons nor I liked the salty film left on the skin after a bath (felt itchy). I mixed some coconut oil in with the salts and water. Actually, it is more oil than water. Three tablespoons water plus four tablespoons salts plus 12 tablespoons coconut oil. The coconut oil is good for the skin anyway and it seems to counter the drying effect of the salts. I found that just mixing the salts and oil did not dissolve the salts, so I needed to add some water. I apply this liberally on the skin and it soaks in plus leaves the skin smooth and soft. Adjust the quantity of salts to your liking.

High phenolic foods, chemical additives, and enzymes

Naturally occuring salicylates and other phenols do not exist in the same intensity in all foods. Members on the Feingold program point out some studies rank foods by the quantity of phenols present in a food as very low, low, medium, high, and very high. These are not absolute values or correspond with exact toxicities or reactions to the foods. Rather it is included only as a guide. The Feingold literature also notes that salicylates are cumulative in the body, and may only be processed out at a certain rate. So, if you consume more than the body can process out, you get a reaction.

As for other chemicals, even small amounts of colorings, sulfites, nitrates, or other compounds may cause a negative reaction, which indicates some sort of pharmacological effect also. For those sensitive to phenols, a strong broad-spectrum enzyme product may help somewhat with phenolic foods. Several parents found they could give low quantities of some phenols with enzymes, but needed to keep track of the total phenol load for the day, or week. Besides facilitating the general breakdown of foods, enzymes may be helping by possibly releasing and enhancing the absorption of more sulfate, magnesium, and molybdenum from foods which are helpful in processing phenols.

In April 2002, No-Fenol became available. It is a very unique enzyme mixture just for assisting with the digestion of highly phenolic foods. No-Fenol performed very well in months of preliminary tests with phenol-sensitive children. Since its release, it continues to give excellent results with actual foods (fruits and vegetables). It's ability to facilitate the digestion of phenolic artificial compounds (colorings, flavorings, etc.) is somewhat limited, and varies greatly by individual.

The exact reason No-Fenol helps is not precisely understood. The phenol processing, sulfation, and detoxification pathways are rather complex. It may have more to do with the specific structure of these phenols than anything else. The research literature indicates that some phenolic compounds are modified by the addition of carbohydrate groups to their structures, which may inhibit their crossing into cells and being metabolized properly. A current hypothesis for why No-Fenol helps may be because the enzymes in this product are able to remove certain carbohydrate groups, or otherwise modify the physical structure, thus allowing healthful digestion and processing in the body.

Some individuals sensitive to phenolic or salicylate compounds are also sensitive to the fruit-derived enzymes (bromelain, papain, actinidin). It may be the presence of some phenols from the fruits linger in the enzymes; or it may be because fruit-derived enzymes tend to high a higher percentage of sulfur bearing amino acids than other enzymes. Or something else entirely different. Enzymedica is one of several companies that makes enzyme products without fruit derived enzymes or fillers. No-Fenol also has no fruit-derived enzymes. Just being able to add foods containing a little bit of fruit can greatly expand your menu. Children like fruit for color as well as for taste.

Fruits and vegetables are very beneficial in maintaining good health. Flavonoids, beta-carotenes, and other phenolic compounds have been specifically identified as important in preventing an number of illnesses, such as various cancers, and have an important role as effective natural antioxidants. So, being able consume these in whole food sources may have substantial benefits later in life as well as now.

Food chemicals and multiple chemical sensitivities

Other chemicals occuring in food may cause problems for individuals sensitive to them. Many of additives are not only harmful in themselves, but they can rob the body of important nutrients like zinc (Ward 1990; 1997). Some common sensitivities are to:

- benzoates (sodium benzoate, benzoic acid)
- calcium propionate
- MSG/HVP monosodium glutamate or hydrolyzed vegetable protein, flavor enhancers
- artificial and synthetic sweetners
- gums (xanthum, guar, etc.)
- nitrates/nitrites
- sulfites/sulfiting agents, corn syrup
- amines/histamine promoting foods

A note about corn syrup: the manufacturing proces of making corn syrup involves sulfur dioxide. A small amount of sulfite residues linger in the final product. Someone may be quite tolerant of actual corn but is intolerant of corn syrup, or large quantities of corn syrup. Highly processed syrups also affect blood sugar levels quicker (see Hypoglycemia page 277).

This further illustrates the importance of carefully evaluating the different components of food and not jumping to conclusions as to the cause of any reactions. With bread, many commercially made breads have artificial preservatives in the pan sprays used to bake the bread in, or in the oils used, or added to the bread to extend shelf life. The problem may be the chemical additives, and not gluten, for example. Bread made without these additives may be well-tolerated. Then, if the bread is still not tolerated, it might be a gluten sensitivity (and would that be the carbohydrate or protein portion?); or perhaps the starch is feeding a yeast or bacteria problem which is causing the reactions? Or the person has an injured gut and the starch or protein digesting enzymes in the gut lining are not sufficiently available. Each of these reasons has a different path to resolve the problem.

Fragrances, pollens, cleaning supplies, fabric treatments, and a host of other environmental chemicals, both occuring naturally and in products we use, can cause adverse reactions if you are sensitive to them. Some chemicals can inactivate or suppress the action of digestive enzymes (e.g. Sastry and Gupta 1978), or be very harsh on the gut lining. Food amines can affect overall histamine levels which can directly impact neurotransmitters. Giving a child with problems real nutrition and additive-free food may not be just a nice alternative, but a basic necessity. Enzymes can free up nutrition from food, however nutrient-rich foods need to be supplied to begin with. Enzymes, whether present in the food eaten, made by your body, or added as a supplement, enable your body to use nutrients in the food you eat. But they cannot convert non-nutrients into healthful nutrients.

Progress check

By this time, we had discovered a number of well-founded, highly researched, scientifically proven reasons why there was so much success with these enzymes. Although we did not know it in the beginning, we now had located very rational explanations why so many people starting with different symptoms and backgrounds were seeing success. Digestive enzymes can support improvement in multiple areas at once. It was logical and reasonable. Enzyme therapy was proving to be a very beneficial and stable alternative. More people were encouraged and decided to try it. You did not need expensive, difficult testing procedures to start. It was affordable. It was practical.

It was wonderful!

CHAPTER 18

The Total Load
– Elephants and Canaries

Some people may harbor excessive amounts of assorted toxins in their body acquired from a number of places. Toxins may come from environmental substances as well as internal sources (such as by-products from bacteria, yeast, maldigestion, malfunctioning pathways, etc.). There are an assortment of ways, methods, even entire books, on how to detoxify the body of particular compounds. Keep in mind that something that one person's body may respond to as a 'toxin' may not be a 'toxin' to someone else.

Some people are just more biologically sensitive to certain substances whereas others are more resistant. This is a combination of genetics and environment. The amount of mercury, arsenic, table salt, or whatever it takes to make someone ill depends on the unique biochemistry and health of each individual. How sensitive they are versus how much their system is required to handle at a given time.

When evaluating a substance, you need to look at the safety of each individual ingredient used as well as the entire product as a whole. Two very different things. The main compound may be tested thoroughly and deemed safe, but the other substances in the complete product mixture may make it hazardous. It is noteworthy that many products do carry specific warnings of side-effects, best use practices, and safe handling instructions. However, these may be disregarded or not fully understood by the person using the product(s).

Even if toxins are not the root cause of a neuro-immune condition, they may at least be a contributing factor in bringing down the avalanche of effects that sends a person's system into dysfunction. Toxin-induced immune suppression may be just the window of opportunity that a pathogen needs to become a problem. An ear infection might get way out of hand, then antibiotics are introduced, leading to yeast overgrowth, and then the spiral down sets in.

The good news is that as you mend one area, this will also have a positive influence on the other areas. Much of the damage can be reversed for many! The body is a dynamic process. Healing, growth, and development are a natural process.

The weight of the heavy metal problem

Heavy metals, as well as other chemicals, are an important part of many bits of modern technology we may take for granted, occuring in many places you may not suspect – in various foods, products we use, and the environment. However, heavy metals are also well-documented as causing neurological damage and dysfunction in susceptible individuals. Possible neuro-toxins include lead, cadmium, antimony, arsenic, mercury, and aluminum among others. (e.g. Landrigan, Graham, and Thomas 1993; Boris and Mandel 1994; Trocho *et al* 1998; more in reference sections).

There is a list of conditions metal toxicity can create – the same list of dysbiosis, sensitivities, autoimmune problems, gastrointestinal and neurological calamities as given in other sections of this book. Surprised?! The list also contains my old chums migraines and sensory dysfunctions. Toxins can affect sulfation pathways, enzymes, fatty acids, brain function, glucose levels, serotonin production, mineral balance, amino acids, and many others.

Having an impaired ability to detoxify adverse chemicals is no help at all. Our bodies should be able to process out a certain amount of these toxins regularly. The cause of a detoxification limitation might be due to an overburdened liver, a damaged gut, an immune system spread too thin, or some genetic factor. This may explain why only some children or adults have adverse reactions to certain substances.

The Pfeiffer Treatment Center has found that a particular protein known as metallothionein (MT) may be dysfunctional in many with childhood neurological conditions. What causes this dysfunction is not known, but thought to be the result of other physiological events. One of the primary functions of MT proteins is to detoxify the body, particularly of heavy metals. Other functions of MT proteins are direct involvment in the development and maturation of the nerves, the brain, and gastrointestinal tract. The timing of any 'environmental insults' to the child is critically important in producing a biological response leading to a neurological dysfunction. These biological systems may have developed sufficiently by the age of three so that environmental toxins no longer result in debilitating conditions. This is consistent with the observation that children rarely 'become' autistic after three years old (although they may display attention deficit traits, milder sensory issues, etc.). More reading on this at *www.hriptc.org*.

Pfeiffer also identified that most of their patients with autism conditions and even some with AD(H)D also have very low zinc levels. Zinc is necessary to intestinal lining integrity and essential for serotonin production (good sleep and even moods). Zinc deficiency results in decreased food intake, reduced immune function, and depressed growth. Low zinc levels make children susceptible to infections and illness, so providing zinc may also do away with repeated ear infections and the associated antibiotics. Giving zinc may be very helpful in increasing zinc levels, reducing harmful elevated copper levels, and re-establishing good MT function. With MT functioning again, the body can then remove toxins appropriately.

There are many different procedures and products used for different types of detoxification. Research the different methods to determine which is best for your situation. An immediate course of action is remove or avoid any current sources of further contamination, whatever they may be for you. Epsom salts may help. Since the liver is a great detoxifiying organ in the body, anything supporting liver health tends to assist detox (i.e. milk thistle is rather effective).

As another side-note, Dana's son improved dramatically in many areas after many months of detox therapy. He could even now have gluten with Peptizyde enzymes, and needed less enzymes overall.

Dealing with the overall chemical intake versus detoxification issue may resolve a host of food and chemical sensitivities, and the adverse reactions they cause.

The elephant and the canary

An appropriate symbol of environmental illness and multiple chemical sensitivity may be the small canary bird. Miners used to take canaries with them down into the mines because these very sensitive birds would be overcome by relatively small amounts of harmful or oxygen-deficient gases. Seeing the birds falter, collapse, or die served as a warning of the danger building in the mines, allowing the miners to clear out before the gases affected them too. Miners now use special safety detection devices and instruments instead of canaries.

The current escalation of people diagnosed with autism, attention deficit, allergies, yeast, autoimmune, and other related conditions may be the little canaries of our communities. They are reacting to their environment, taking in a greater toxic and chemical overload than they are equipped to handle.

This is also reminiscent of Pottinger's cats. The first generation with a full supply of enzymes was in complete health. The group on the enzyme-rich diet remained healthy while those on the enzyme deficient diets progressively got much worse with each generation developing more devastating health problems earlier and earlier in life. Adding chemicals and potential toxins from food and the environment on top of fragile health may be like giving a person sledding downhill a huge shove.

Perhaps our current situation may be compared to seeing ever more canaries, getting ever smaller, ever weaker, ever more sensitive.

Now consider the story of the six blind men who did not know anything about elephants. One day, they encountered an elephant and began to feel all around it's body trying to make sense out of it. The first blind man felt the elephant's trunk and declared the elephant was like a snake. The second blind man put his arms around the elephant's massive legs, and said it was like a tree trunk. The third blind man handled the tail, and announced an elephant was like a rope. The fourth man reached high and wide across the elephant's broad side and proclaimed it was like a wall. The fifth blind man felt the breezy ears flapping back and forth, and assured the others this animal was definitely like a fan. The sixth man felt all around the long pointy tusk and was emphatic this creature must obviously be like a powerful spear.

Each man was correct in his assessment of the elephant from the information he had, and yet each did not grasp the full meaning of what an elephant was. Even considering two or three of these views together does not tell the entire story. This appears to be the situation with many conditions. Just as it is hard to see exactly where the body ends and the legs or trunk begin, the gradual flow of one aspect of a condition is muddled in with others. All the parts are connected together making up the complete animal we are dealing with, It is important to keep in mind that all the parts, like dysbiosis, leaky gut, nutrient deficiencies, food intolerances, etc. will rarely exist in isolation. Like a stack of dominoes, they each bump into and influence the next one. So, athough you may be examining one part in intricate detail, this perspective needs to be balanced in view of the whole.

Because of this, it is important to listen to what specialists tell you what you *can* do for you or your child, and think twice about what they tell you that you *can't do.* Think of six blindfolded doctors doing a physical exam on the elephant and you will have a better appreciation for how specialists can help you. Take and use what each source can help with, leave the rest, and move on. Practitioners are all individual humans each trained in a very narrow, and often very different, specialty. The one who says there is nothing you can do may very well only be trained in turtles or penguins.

Each person has a unique blend of the parts, or there may even be subgroups. One person's 'elephant' may have huge long tusks (dysbiosis) while someone else has very small or no tusks. One child may have to carry the weight of a huge load of accumulated nutrient deficiencies or environmental chemicals, but with another person, this is a minor contribution. How much any person's immune system, neurological system, and detoxification system can handle is a matter of degree. The symptoms or behaviors seen may depend on where the dysfunctions are greatest in the body, which part takes the most injury, how that individual expresses their load, and responds to it. Just as different autoimmune conditions arise depending on where in the body the immune complexes settle. It is a matter of how large your particular elephant is, and how large it becomes.

This concept of a progressive condition is not new – that is exactly what the concept of a spectrum is based on. A child may express a form of autism where an adult may develop an autoimmune disease such as fibromyalgia or multiple sclerosis (cumulative effects over a lifetime). This is not an exact description, only a way to describe the relationships. This is also the concept my neurologist tried to explain in the very beginning when she directed us to <u>do what we can to reduce the total load.</u>

The connection to keep in mind is, really, how big is the elephant your little canary is trying to hold up?

CHAPTER 19

Enzymes and Restrictive Diets

One of the best things about being in the Enzymes and Autism group is that it is so positive. It provides support and ideas for making progress. Of course, this is much easier when you have a steady influx of terrific gains, children recovering from tremendous difficulties, and happy families. One of the underlying premises is that a person's diet is not the main issue even though food, eating, and digestion are. It is about respecting the individual parent's judgment. You are not criticized or pressured to change whatever type of menu you do or do not choose to have. Many of the members were still very much on a casein-free, gluten-free, soy-free, or other food-free diet. Like Patti:

> My son changed shortly after his first birthday. He became aggressive, hitting me, and pinching me almost out of the blue! No matter what we did, as he grew, so did his aggression. I literally could not leave him alone because of his attacks on his older sister and mauling our sweet cat. He also started with sensory issues, insisting his food, drink, and clothes all be warm when they touched his body. No more hugs or kisses. He stopped giggling and playing as before. He became very grouchy and nasty by the time he was two years old, obsessing over everything. It was so sad. Trying to discipline him was a nightmare. The child psychologist said he was 'strong willed'

(understatement). We did not have a clue as to what was wrong with him as his speech was advanced for his age, he potty trained at exactly two years old, and always maintained eye contact.

Someone finally mentioned trying food allergy testing, which we did. We found out he was 'allergic' to soy, egg, dairy, gluten, and other things. That was when we took him off dairy first and then gluten, and slowly most other common foods. My little one got about 50 percent better on the diet but still not anywhere near good. A friend gently directed me to a book on autism and we started to make the connection. We also got lucky and found a cooperative doctor.

I heard chatter about enzymes on one of my listserve Internet groups and was guided to the enzyme group. I read a lot and asked questions. Cindy's story convinced me to try the new enzymes. Now my son no longer has ANY of his behaviors. No tantrums. No more obsessions. We no longer have to warm his clothes, he transitions wonderfully, gets himself dressed for school and bed, which used to be a major fight. NO MORE HITTING (a major big deal after two years)!

And one of the biggest changes is he wakes up happy. He also plays with his older sister wonderfully now. He loves to be hugged and kissed (yay) and loves to have his head rubbed. He is suddenly interested in writing. I cannot say enough about how it has changed our lives. For the first time I now have hope for my son, I am not scared he will grow up and hurt people, which was a big fear of mine.

Most people on any type of restrictive diet still see good improvement with the addition of certain enzymes. At times, many parents find that after you successfully remove the casein and gluten foods 100 percent, your child 'becomes' reactive to corn…and then soy…and then phenols…and then eggs…and then… Although this may not be the case for all, it is for many. Some people find that corn, soy, rice, or some other food is as big a problem as casein or gluten foods. Also supporting this is that when enzymes are taken across the board for *all* foods at *all* times, some see a significant synergistic effect which is greater than when taking enzymes either part-time, or for only certain food groups – the Happy Child Effect. Enzymes tend to address multiple issues and foods in one pass, whereas restricting foods can be a tedious, unpleasant, and expensive process. This may contribute to the very high success rate with enzymes.

Many people with autism or neurological conditions have found a number of restrictive diet types to be helpful. These include yeast diets, rotation diets, Feingold and Failsafe programs, Specific Carbohydrate Diet, casein-free/gluten-free diet, and high protein/ low carbohydrate diets. All of these attempt to address some aspect of gastrointestinal health already described. Most diets are concerned with addressing poor digestion or insufficiently digested foods of some sort. Since the versatile digestive enzyme is very effective in dealing with the same gut issues, it makes complete logical sense that enzymes can assist with any or all of these diet types, and perhaps even eliminate the need for a diet completely.

One of the biggest problems with attempting to eliminate 100 percent casein, grains, corn, sugar, or other common foods is that it is practically impossible. A major problem is the many hidden sources of possible contamination. It is very hard to keep up with companies constantly changing their production methods and ingredients. We also have the problem of going places that simply are not that careful about contamination. Another source of difficulty comes from others offering 'unacceptable' snacks, like during school, from relatives, or at children's activities. Just having some 'insurance' for contamination or hidden ingredients, or being able to have occasional planned infractions can greatly improve your quality of life. This backup can remove some of the pressure and stress from parents trying to micro-manage every aspect of their child's life.

Digestive enzymes can enable you to actually be on a restrictive diet to a greater extent than you can be otherwise by trying to target and eliminate food yourself. It provides a realistic, workable solution. Whether someone obtains the best results from doing just a restrictive diet, using just enzymes, a combination of both avenues, or neither alternative, the individual or specific family is the best one to determine this. Each situation is different and results can vary for any particular individual. For some people, they may still need to, or want to, avoid certain foods or chemicals even with enzymes.

Cost of enzymes

Remember that different enzymes work on different foods and a particular formulation may target a particular need. So what is 'more effective' for one person may be different than what it is 'most effective' for another. You may also need to give many more capsules of one brand than another.

Some parents having given enzymes for more than two months said their monthly bill for supplements and therapies is decreasing even with the additional cost of the enzymes. This correlated to length of time on enzymes and to what extent they remained on a restrictive diet. Specific reasons include:

- Do not have to buy as much special food. It is usually much cheaper to give enzymes even some of the time than to buy the special food all the time. You can save on the expense of the food as well as any shipping charges.

- Can take advantage of sales and specials at the local grocers.

- The child or family can eat food presented at an activity or relative's house instead of bringing their own food everywhere. Now this was nice. If we were out shopping and got hungry because the errands were taking longer than expected, we could stop and get a bagel, pretzel, drink, sandwich, or anything else at a snack shop.

- Do not have as many doctor visits. This refers to the constant visits for infections, ailments, congestion, rashes, etc. where there was a need to have someone look at the child and perhaps get a short-term prescription. Children appeared to be healthier in general and there were less visits of this sort. I know I was going to the doctor at least once a month for something. Infection, cold, pains here or there. I have not been to the medical center in months.

- A healthier, more attentive child means that other activities such as behavioral, speech, and auditory therapy may be much more effective. Many parents said their children were progressing much quicker and so they would be finishing with therapy sooner than expected. Or the child needed much less intense therapy each week. A few said their yeast treatments were also more effective.

- Do not need to buy as many supplements, or perhaps medications. With constant food removals, you may need to keep adding more and more supplements to correct the nutritional imbalance in the body. Several families using enzymes regularly for over two months were able to reduce the number of supplements they were giving, and in a few cases even reduce medications. We have gone from around 20 supplements to five while another mom reported decreasing from over a dozen to three, and others reported just generally not renewing supplements as their supply ran out with no regression. This makes sense because you are getting better nutrition from food, and using it more efficiently. Besides the special food, this is the other major area for cost savings that greatly off-set the cost of the enzymes (at times saving thousands of dollars in total per year).

Convenience of enzymes

Most parents say the use of enzymes has been far more convenient than a restrictive diet. The main reasons given are:

- We have the obvious benefit of being able to eat more foods in more places with less hassle.

- Shopping is easier. You can buy from more places and buy a wider variety of things.

- A particular benefit is noted by those parents with older or school-age children. Several people found a diet very difficult to maintain with an older child because of the child's non-compliance, having a broader range of activities, and attending school. The enzymes were easier to manage by the child themselves which is particularly helpful when the child was at school or on his or her own.

  For one, it makes it easier to allow the child to eat a school lunch and participate in activities involving food. Another is that many children this age begin to want to fit in with their friends. Improving social skills and interaction is usually a goal anyway. Removing food restrictions as a major obstacle makes it easier for the socially-challenged person to manage social situations. Some children are old enough to recognize that they feel much better on enzymes whether foods are involved or not, so they are very diligent about taking their enzymes themselves.

• A few parents have mentioned how much better and easier it was to take enzymes when traveling, visiting friends or relatives, and going to other locations.

• Using enzymes may more efficiently break up any offending peptides, sugars, or other foods rather than trying to 'guess' what to eliminate without them. Enzymes greatly reduced the stress and frustration level of parents trying to monitor everything their child came in contact with along with figuring out which foods continued to be 'safe' and which were not. Also, trying to rebalance the person's nutritional needs with supplements was noted as being difficult, at times dangerous, tiring, and expensive.

How long before seeing benefits

Most people say they see results they consider positive in the first week (at least 50 percent). By the second to third week, this increases to about 80 percent. If positive results are not seen by the third or fourth week, it is recommended to consult with others and see if they can assist with the problem. Or call the manufacturer or ask a health practitioner for ideas. If no results are seen after this, you may want to try a different enzyme product before giving up on enzymes completely. People do react differently to different formulations.

Many parents report positive improvement in the child and/or no negative reactions from the very first dose of enzymes. If you do not see improvement or results into the second or third month, then it is debatable whether enzymes will produce improvement with further use. It would be up to the parent's discretion to use enzymes past that amount of time or not.

More time may be needed to heal a leaky gut. Older children may need a little longer as well. Looking into a yeast issue or detoxification program would be a good next step as these two issues seem to mask the observable benefits of enzymes in some. There may also be a food or two in particular you need to eliminate.

Effect of age

Ever since I first started reading about treatments for autism, particularly with diets, I read that you must start therapies as young as possible or the person has little chance of improvement. The window of opportunity was up to age five or six, and then it supposedly closes for the most part. Trends with detoxification therapy and the casein-free, gluten-free restrictive diet both showed that younger children showed much more improvement than older children. Well, because my boys were already seven and nine years old, I promptly disregarded that. Although starting as young as possible is preferable, I did not have the option of 'starting younger' anymore and fully intended to keep looking for something to relieve the pain, if possible.

I was encouraged by what I had learned in college, which was that there were two primary times during your life when the brain prunes its neurons. One is around age five or six years, and the other major time is at puberty. Besides, people learn all their life if they want to, so a chance of some kind of improvement is always possible. Your body is constantly growing, healing, and changing.

There is excellent news for those with 'older' children, meaning around age six and up. Older children through the age of ten were just as likely to show improvement with Houston enzymes (the ones tracked) as younger children were. Progress in older children was just as significant and just as immediate as with younger children. This trend appeared whether the child was on a restrictive diet or not.

For those over ten years old, we have not collected as much information simply because fewer individuals in that age group happened to come along and offer comments (this is the same reason there are only a few cases involving initially low-functioning individuals). Of the cases reported, children between 10 and 18 years old along with adults themselves fall fairly evenly into four groups of improvement within the first 4 to 6 weeks:

- those showing no change
- those making slow, marginal yet progressive improvement
- those making moderate to good improvement, and
- those making the amazing, impressive gains

Overall, teenagers and adults are very likely to show some sort of improvement but the degree may be lower initially and may simply take a bit longer. However, good improvements do come for many, particularly if enzymes are tried for at least two to four months. Another reason that older children or adults may take longer is because they have learned various coping skills and patterns along the way. It may take longer for them to adapt to feeling better, and learn newer and different ways of acting and reacting.

Why do older children as well as adults generally show greater positive results with enzymes than with restrictive diets? One reason could be that older children and adults have simply had more time to develop far more extensive gastrointestinal problems, food intolerances, and metabolic problems than younger children. So if a younger child is intolerant of three foods and an older child is intolerant of 17 foods, eliminating just gluten and casein (or any one or two foods) may simply not be enough to achieve the higher levels of improvement. You would not see as many positive results by only removing a couple food types with the older person. Even if many foods are removed, this may not enhance gut healing.

Using enzymes breaks down all foods at the same time. In our example, both the older person and the younger child would have all their problem foods addressed by enzymes at the same time. Thus, more age groups show greater improvements immediately with enzymes. Enzymes may also provide pro-active gut-healing benefits as well as other health improvements addressing a multitude of biological dysfunctions at the same time.

Another factor could be that parents have more absolute control over the diet and environment of younger children, whereas older children are more independent and have more opportunities for exposure to off-diet items. Parents of elementary school-age children find their children are far more willing to comply with taking enzymes than adhere to a restrictive diet. School-age children and teens tended to be surprisingly cooperative in taking enzymes, even reminding their parents and teachers. Older children could determine they just felt better taking enzymes. The main reasons given by the children themselves were wanting to participate in social activities like others, have school lunches, feel better, and just eat something they enjoyed.

Can enzymes replace a restrictive diet?

Enzymes do not simply replace a restrictive diet. Enzymes operate in a different way and contribute to many other types of healing. Enzymes may be much better than just doing a restrictive diet because they can support intestinal health pro-actively. It is not a one-to-one even substitute. Enzymes do a great deal more.

From a practical standpoint, particular enzymes may quite possibly eliminate the need for the casein-free, gluten-free or other restrictive diet. For many, Peptizyde is able to accomplish the task of eliminating the casein, gluten, and other suspected peptides from the body and preventing possible adverse effects, even when eating casein and gluten liberally. The majority of people using the Peptizyde enzymes have been able to reintroduce casein and gluten, as well as other foods, if they so desired. Many people have successfully left a casein-free, gluten-free diet seeing even greater improvement. These families report no problems or regression after more than two years. The main exception is people with celiac disease should still eliminate gluten.

This experience can be supported with lab tests. One doctor tested a patient with autism showing good improvement using Peptizyde. The child was given mozzarella, gluten, and ice cream with the enzymes. Urinary peptides indicated an 'insignificant' amount of casein peptides and no gluten peptides. The child was not on a casein-free, gluten-free diet. The insignificant fraction may be due to the basal level of the test or other unknown factors that normally occur.

It is important to remember that only a few enzyme products are designed specifically for this particular protein issue, starting with the release of Peptizyde in April 2001. These are the only products effective enough to use with dairy and grains regularly. Other products were either not designed for this or were evaluated as not effective enough to use all the time for this purpose by their own manufacturers. For other specific needs, look for enzyme products appropriately formulated for targeted purposes. Different diets are geared to accomplish different things. Enzymes may be able to replace or assist with some aspects of these diets, whereas other aspects won't be affected at all (i.e. additives targeted by Feingold or Failsafe programs).

What about substances that enter through the skin? There is no conclusive information one way or the other how taking enzymes works in this situation on specific proteins like casein or gluten, or with chemicals such as artificial colorings. With food proteins, most people who have added foods back in with enzymes have not detected or mentioned having any problem with non-food products such as lotions, shampoos, craft supplies, etc., but it is something you may find you still want to avoid. Since proteases are known to enter the bloodstream and circulate about breaking proteins and immune complexes down, perhaps after taking enzymes a while there are enough proteases in circulation to break down those compounds should they enter the body. Amylase is known to break down sugars in the bloodstream while proteases are known to enter the blood for other types of cleansing and detoxification.

With chemicals like artificial colorings, fragrances, preservatives, etc., enzymes have not proven to be that effective so sensitive individuals should still avoid these. This makes sense because these compounds have different chemistries and are not true 'foods.'

Having an effective enzyme product for carbohydrates, casein, gluten, or other foods can be a blessing for some parents who do not have care of their child full-time. More than one parent has mentioned that being on a restrictive diet was impossible before enzymes because their child was periodically in the care of another person who was not very cooperative with a particular diet plan. Providing enzymes is usually much easier to implement, and this might just be the vehicle to enabling a child to be on a particular diet overall on a regular basis. It has been critically important for an enzyme product to be developed as an effective alternative or replacement for the casein-free, gluten-free, or other food-free diet for this group who previously had no option to help their children or themselves through no fault of their own. Finally, this essential resource is available.

No-Fenol and special diets

No-Fenol is proving to be a very interesting product. Many people are able to immediately return basic fruits and vegetables back into the diet. It may also help in a very limited way with some artificial compounds. This enzyme blend may be a major help to those on a Feingold-type program, or those who cannot implement such a program because of circumstances or other reasons.

In the Feingold Program, a group of salicylate foods are removed and then reintroduced one at a time, testing for tolerance. The intitial period where both the phenolic additives and salicylates are removed is called Stage One. Once a good response is observed, the individuals may move on to Stage Two, when they reintroduce natural salicylates. It is too early to say if certain enzymes will enable salicylate-sensitive people to add back more of the salicylate-containing foods, but it looks promising (going from Stage One to Stage Two). We do not yet know to what extent enzyme supplements will reduce a reaction to synthetic additives, but the many thousands of families using the Feingold Program will provide valuable feedback as we learn more about enzymes. Artificial additives are not nutritionally necessary and may have long-term detrimental effects on health and neurology.

The Specific Carbohydrate Diet (SCD) is intended for those suffering from severe bowel diseases. It focuses on starving out bacteria and only allowing foods that do not require certain intestinal enzymes which may not be available due to gut injury. This diet has helped some with autism where other restrictive diets have not.

When carbohydrates are not digested and absorbed in the small intestine, they remain as a source of food and fuel for microbes in the small and large intestines. They produce harmful acids and toxins that injure the gut and pollute the body. This starts the cycle where further damage to the gut only promotes further lack of digestion and nutrient absorption, more microbial growth, and an increase in toxins. When their food source is removed, the microbes die-off, harmful by-products are reduced, the gut heals, overall digestion improves, the immune system strengthens, and good health returns. This strategy is similar to yeast diets where foods promoting yeast are eliminated. High protein/low carb diets are similar in that they reduce the load on the pancreas to produce sugar and starch digesting enzymes and alleviate the amount of carbohydrates requiring the enzymes located in the intestinal lining for breakdown.

Recall how many individuals with neurological and intestinal problems were also found to be very low in lactase and disaccharidases.

No-Fenol appears to be accomplishing or contributing much as the SCD and perhaps other diets do with food eliminations. Enzymes such as cellulase and xylanase break down the more complex foods and fiber leaving less for adverse microbes to use. Similar enzymes are commonly given to livestock specifically for this reason. The idea is to maximize the breakdown of the harder to digest fiber in feed and provide more nutrition and energy from the food to the animal (or person). Those starting No-Fenol often noticed a surge in energy and feeling more alert. Giving a range of enzymes with all foods probably contributes to eliminating an excess food source for pathogens as well. Enzymes are also used with animals to reduce the total amount of waste moving through the system. Having more rubbish in the gut can promote harmful pathogens, toxins, and excess ammonia. With No-Fenol, some people noted a decrease in total quantity of waste, not to be confused with actual constipation where the waste is just too condensed and building up. Later, another enzyme product, Ultra-zyme Plus from Thropps Nutrition, turned up showing similar results of 'improved elimination' and 'improved bowel response.' It also includes much more of the fiber digesting enzymes.

Besides reducing the microbes food supply, No-Fenol acts directly as a yeast fighter. Within days of its release, a number of parents said they saw a jump in yeast control after adding No-Fenol. Yeast have outer coatings made of protein and cellulose. No-Fenol contains a lot of cellulase which no doubt contributed to eliminating some yeast. Once the cell wall is broken, other yeast-fighting herbs or medications can get in and more effectively conquer the yeast. The combination of No-Fenol and grapefruit seed extract has proved to be very effective against yeast. Someone on a yeast diet may find products containing high amounts of cellulases and proteases to be a helpful addition.

It is interesting to note that a few people who did not see great improvement on other types of enzymes saw a burst of improvement with No-Fenol, just as some trying other elimination diets did not see improvement until they did the SCD. Yeast flare-ups have been responsible for masking the gains seen with enzymes in several other cases, and No-Fenol enzymes attack yeast, so that would be a consistent explanation (which pleases me no end – I like consistency).

Another reason may be the speculation that these particular enzymes help the body process certain amino acids with phenol structures such as phenylalanine, tyrosine, and tryptophan. Perhaps these are better utilized appropriately in the body.

Should I try enzymes if special diet didn't work?

If someone has attempted a restrictive diet and feels certain they were 100 percent in compliance, and did not see any reaction, positive or negative, they may want to know if giving enzymes will help bring improvements. With the casein-free, gluten-free diet, the information collected so far indicates that for all age groups, you have about a 50 percent chance of seeing improvement with enzymes and a 50 percent chance of again not seeing any change at all. Since parents are advised to give a casein-free, gluten- free diet up to a year, you could also conclude that enzymes should be tried regularly for a year before deciding conclusively. No decisive information is available for other particular carbohydrate diets or protein diets.

Allowing about four to six months would be reasonable for gut healing to take place, which may be a good benchmark. A few parents of low to moderate functioning teens reported that they did not notice improvement until into the second, third, or fourth month. So an older or lower functioning person may need longer than the three to four week typical time frame to see notable progress.

Even with enzymes, there may still be a food or chemical that is not tolerated and clouding the results (examples: amines, sulfites, paraben). Food additives can greatly influence neurology and are out of the direct realm of digestive enzymes.

Remember also that not every person's condition may be heavily influenced by or based in gastrointestinal issues. The person may simply not have a digestion issue at all. Even some gut issues make be fundamentally sensory-based and not involve food content.

The dose makes the poison

What about the argument that giving enzymes with dairy or any other problem food is like giving the antidote along with the poison? Depends on how you choose to look at it. That is a very dramatic statement no doubt intended to convey the urgency to conform to a strict and often expensive path. For many people, this statement is simply not true. What if the child actually has a leaky gut, or inflammation, or environmental toxicity, or some other situation that disrupts digestion in general? Is the solution to withdraw food upon food until you are food-free? The food may not be the basic problem and eliminating it may not contribute much to facilitate healing the gut. There may be a biological breakdown somewhere along the line and a particular food is just a bystander caught in the fallout. It would be better to skip the emotionally-charged banter and look at how something is designated as a 'poison' to begin with.

Almost everything is poisonous in some amount. 'The dose makes the poison' is a standard saying in toxicology. This means that the dose is what makes something poison or not, and *not* just the substance itself. One example is table salt. A tablespoon of table salt can kill a child. Okay, so I say salt is poison to my child, and I need to make sure he never gets one single grain or molecule. Well, if I do accomplish that, the child will die because salt is essential for life. Animals instinctively know to go to a salt lick for salt, and farmers must ensure salt blocks are available for their livestock's health. Salt is essential at one dose, yet toxic at another. Salt has a lot of company in this matter. Although essential for health and life, vitamin D is very toxic if very much at all is consumed. Dishwashing soap is more toxic than many pesticides.

The standard basis for determining toxicity for comparative reasons is known as the lethal dose, or LD. The LD standard is based on a benchmark value where 50 percent of a test population (usually rats or mice) dies from exposure at a given dose. This will be different for each substance, and is known as the LD50 of that substance. The higher the LD50, the less toxic the substance. The lower the number, the more toxic it is. The units on the LD50 are given in weight – the weight of the chemical per the weight of the animal.

The LD50 for Roundup™ herbicide and Abound® fungicide are 5000 mg/kg, while the LD50 of table salt is 3000. Therefore these pesticide chemicals have been found in laboratory feeding studies to be less toxic than the regular salt we consume every day. The LD50 of methyl bromide is 214 mg/kg, which is less toxic than caffeine (LD50 = 192) or nicotine (LD50 = 55). The acute toxicity of caffeine demonstrates that small quantities of such a chemical can be ingested in small quantities daily without adverse effects for most humans. Caffeine can even be helpful for migraines and other uses. Many medicines are fatal at high doses but may be essential for a person to survive or overcome a medical condition. Strychnine and parathion are both deadly toxins with an oral LD50 of 10. Vitamin D has the same oral toxicity of 10. If vitamin D were not exempt from the standard regulatory safety requirements, you would have to wear a protective 'moon suit' to take your morning vitamin pill.

Because table salt has an LD50, meaning that at a certain dose it is toxic, we cook and eat 'poisonous toxins' every single day. Salt is a common and necessary part of our daily diet. Since salt is necessary for life, you can easily see that this 'toxin' is necessary for survival.

Keep in mind that even at these LD50 doses, only *half* of the test population is killed, the other half is not killed. Although animal studies are not an exact equivalent to humans, they are useful as a guide.

Besides the dose, there is the issue of rate and time. Because of these two factors, we can safely take the vitamins and salt we need, and don't collapse from breathing air or drinking water. Let's say a box of salt contains around 48 lethal doses. You eventually feed it to your whole family over time and it is processed normally performing vital functions. Forty-eight people are not killed. Even the few members of your family are not killed by the time the package is used up. How about aspirin? An entire bottle of aspirin may contain one lethal dose, or 100 tablets. If someone takes it all at once, this is poison, and we can expect the worst…maybe. He might be in the 50 percent of the population who is not harmed at the lethal dose. We can also distribute them around town, giving each person only one tablet, and not see ten deaths. We will not even see one death if we eventually use the entire bottle ourselves over time.

Many people feel that if foods are being properly digested with enzymes, then the foods are not 'poison.' They are nutrition. Very valuable nutrition! With a injured or malfunctioning digestive system, any food or supplement could be considered 'poison.' It is better to fix the faulty system, not focus entirely on eliminating foods on end trying to find that elusive offender. There is also the widely acknowledged point that gluten and casein proteins are not potentially hazardous as proteins, nor when completely broken down. The proteins are only a possible problem when digestion happens to be faulty, and the intermediate peptides happen to form, and happen to counter this exact same situation. This is a very standard, well-studied pass into the bloodstream, and happen to cause an unwanted reaction of some type. They are not even harmful to all people with autism.

There is another concern. An antidote neutralizes a poison, or otherwise renders it harmless. However, we have seen time and time again that giving enzymes plus previously troublesome foods may bring immense improvement over just eliminating the food, and even over giving enzymes plus eliminating the food. There were dozens of people by now who had been 100 percent casein-free, gluten-free, and/or other food-free for a year or more and saw improvement by restricting these foods. Then they added enzymes and saw more improvement. Then, if they were using Peptizyde, reintroduced gluten and casein foods, along with other previously restricted foods, and found their children were much better – happier, healthier, better behavior – off the diet plus enzymes than on the restrictive diet.

This phenomenon occurs even when the parent was very diligent about providing a well-balanced diet and many supplements. So, this appears not to be a break-even exchange. The enzymes plus food is obviously supplying things that the body truly needed and by withholding this, the body was actually staying in a suboptimal or unhealthier condition for some people. In these cases, it appears that food elimination was more detrimental to the body than giving the enzymes plus food.

So, based on actual science, the months of results of many parents, I cannot see how describing 'food as poison' is at all a universal truth. These enzymes provide newer information and a newer alternative that was not available previously. I sincerely hope no one would be discouraged from considering other, possibly better, alternatives based on this older extremist argument.

Opiate-receptor mechanics

Some sources may say using enzymes is not an effective method of controlling all the potentially harmful peptides or compounds possibly produced by consuming casein, gluten, or other foods when used as an alternative for a casein-free, gluten-free diet. Many of these sources may only be considering enzyme products in the general sense and not specific products formulated for specific needs.

There is now much evidence from people eating dairy and grains liberally and regularly with Peptizyde. The vast majority of these people say that the enzymes they use are very effective, and the effectiveness has lasted over a substantially long period of time (over two years now). Because my sons and I developed migraines within three hours after consuming casein in the beginning, I was rather confident I would know if the casein was not being broken down enough. Others families say the same thing. Many could identify definite physical or behavioral 'markers' for a variety of foods. After all, it is seeing problematic behaviors or physical reactions that lead you to eliminate a particular food in the first place.

The line of reasoning that states enzymes cannot break down all the suspected peptides so therefore they should not be relied upon exclusively, fails to account for how opiate-receptor mechanisms work. If we are operating under the assumption that these potentially harmful peptides, called exorphins, are interacting with opiate receptors, then we have to go by what is known concerning receptor mechanisms.

First, it is impossible to eliminate all opiate-like exorphins in the body, or from the diet for that matter. '100 percent casein-free, gluten-free' does not necessarily translate into '100 percent peptide-free,' as many other food proteins are potential sources of exorphin peptides, including the blood itself. The body needs some exorphins and will generate them. So, being 100 percent casein-free, gluten-free rarely, if ever, removes all peptides to begin with. Since people routinely say they are 'immensely' improved at times by just an elimination diet, then this validates that you do not need to eliminate every single molecule to achieve immense improvement. The fact that many people see even better improvement off a casein-free, gluten-free diet with enzymes (at least with specialty enzymes like Peptizyde) validates that the enzymes may do a more thorough job on their own than just eliminating foods.

Next, must one eliminate 100 percent of all exorphin peptides from the body, whether by diet, enzymes, diet plus enzymes, or some other means? Can one molecule of dairy or wheat cause problems in children with autism? In all honestly, this is highly doubtful. If we base the problem on opiate-like reactions with receptors in the gut, brain, and elsewhere, then we must also look at how those reactions actually work. The peptide must be physically bound to the receptor molecule's binding site. This binding can trigger the nerve reaction that might produce the negative behavior. One very well-characterized aspect of these receptors is that a certain number of opiate receptors must be occupied to have a receptor-mediated event occur (a percentage). The basis for most modern pharmacology is the recognition that medications, hormones, and peptides work by binding to receptors on cells and must occupy a certain percentage of those receptors to elicit an effect. So, it is based on a percentage and not each individual molecule, or even one exact number of molecules.

For example, if one is in pain, then morphine may be prescribed. However, one molecule of morphine rarely, if ever, causes a decrease in pain, nor does two molecules. You may need 5 milligrams of morphine before you feel less pain. It took that much morphine to bind to and activate the percentage of opiate receptors needed to produce the effect. Any less was not effective, but it was still in your system. Pain will not be decreased until a certain amount of morphine is bound to a certain percentage of morphine receptors.

Let's say there are 100 opiate receptors available, and all their binding sites are empty. In this example, we will say it takes 25 percent occupancy of receptors to trigger a particular reaction. We will say the reaction is 'constipation.' Now, ten exorphin peptides come along which proceed to bind to ten of the opiate receptors. We now have 10 percent occupancy of available receptors. At 10 percent occupancy, no constipation occurs because it is below the threshold level of occupied receptors needed to trigger that particular event. You may even have a constant flux of a few peptides leaving the receptors, while different new ones bind. But you remain below the threshold level so there is no 'event.'

Then 15 more peptides come along which bind to the receptors. Now we have 25 percent occupancy and constipation occurs. If even more peptides come in, like about 15, we will have 40 percent occupancy. However, the event is already in action, so possibly the only additional event to occur is possibly a prolongation of the constipation, because there are more occupied receptors that will have to become unoccupied for the receptor event to stop.

The point is this: often, all that is needed is a decrease of exorphin peptides to a certain threshold level to observe positive effects, not eradicating every molecule. Diet and/or appropriate enzymes may both accomplish the action of lowering the exorphin peptide level.

There is *always* a critical amount of substance needed to be bound to the receptor population to elicit an effect. That percentage may be 10 percent or it might be 80 percent, it is going to vary by person and circumstances. The particular percentage will vary depending upon a number of complex factors that are beyond the scope of this discussion. It usually has to do with:

- individual physiology
- the nature of the substance
- the quantity of the substance
- how often the individual is exposed to the substance
- the ability of the individual to eliminate the substance
- and, the actual dose of the substance at any given time

Most medications working through receptor systems have dose-response studies attached. The response is dependent on the dose, and is a well-known, well-characterized system. Our bodies produce a certain amount of opiate substances naturally, so are continually dealing with this situation anyway whether you eat certain foods or not. Don't forget peptides have many beneficial functions as well. We may have substantial health and nerve problems if we did eliminate every exorphin peptide. Many essential neurotransmitters operate in the same way.

Tolerance and sensitivity

Another factor that comes into play is the mechanics of tolerance and sensitivity. Many parents have noted variable responses to 'problematic' foods. Some kids can eat more wheat or dairy with no problem, but others seem to be set off by the tiniest bit. Some parents notice that this tolerance appears to vary over time, whereas others do not see any change in tolerance levels. The body may adapt to many biochemical changes and this is no different.

In days gone by, one technique of poisoning thy enemy was to intentionally slowly consume arsenic over time and thus you would develop a 'tolerance' for arsenic. The body adapts. Then you would have the person you detest over to dinner or for a drink prepared with arsenic. You would have the same food or drink but you would not suffer whereas your foe would keel over.

The point here demonstrates that the threshold limit may change over time, by circumstances, or exposure to a particular substance. We also have the inherent genetic sensitivity to a substance to factor in. And, the body may compensate by creating more receptors or less receptors over time. Again, we are dealing with a percentage at any given time and not an exact number. The situation gets even more complex because the activation state or affinity may vary as well.

To understand 'affinity,' think of two people shaking hands. The grip of one hand on the other may be loose, medium, or tight. The grip may be looser or tighter over time and this results in a nerve response of more or less, just as the recipient of a handshake will feel the change in grip pressure.

A balance develops between substance concentration and receptor occupancy in everyone. When this balance is disturbed, there may be a physiological or readjustment effect. One of the possible events is decreased pain sensitivity (analgesia). As the relevant peptides are reduced, these unoccupied receptors stop signaling to the cell to which they are attached, and the result is increased sensitivity to pain or other sensory input. This is the basis of the 'increased awareness' or 'increased sensitivity' when starting enzymes. It is positive overall, but at the same time the person needs to get used to the new level of and types of sensory input. This compensatory reaction is part of everyday life. Endorphins and dopamine are made naturally in the body and have very useful, necessary, and essential functions.

It is also interesting that many studies have been done to determine if patients using morphine for real pain, such as with cancer or other painful illnesses, do not become 'addicted' to it. Just as salt and vitamins are essential to proper health in the appropriate amounts, so might be casein, gluten, corn, or many other substances once the mechanism is functioning well. So, if enzymes contribute to healing a leaky gut, and then break down foods appropriately, any negative peptide effect or immune system response can be eliminated, and the foods resume their proper role in enhancing health and well-being. This may explain why many people see the best results when using enzymes and not on certain restrictive diets. Many times a certain amount of something is needed for proper health or even life itself, and is only problematic outside a given range.

As a follow-up, after Patti wrote me, she stayed strictly casein-free, gluten-free, and soy-free for six months with enzymes. She has very recently reintroduced all of those foods regularly and reports her son is doing even better than before, showing another jump in significant gains.

Even if someone is 'craving' dairy, grains, or other foods, there are many possible reasons. It may be the body trying to self-correct in an effort to raise serotonin levels, or relieve a nutrient deficiency. Those are traditional comfort foods. If a person is in pain or discomfort for any reason, this may be a mechanism the body employs deal with it. The person may also be responding to some sensory aspect of those foods and nothing else (color, texture, smell, etc).

Doing a food challenge or reintroducing foods

What if you would like to 'challenge' a food previously restricted or would like to add back a previously eliminated food? The current recommendation is to give a small, known quantity of the food to be reintroduced and carefully observe changes over several days, up to about two weeks. This is because some reactions are cumulative or may not show up after just the first day. Do one food at a time. Having a known amount of food is best so you will be able to see if the enzymes plus food are working.

Example: give one bowl of cereal, not the entire box or letting the child eat until he chooses to stop. If the quantity of food is not fixed, you will not be able to judge how much food in relation to enzymes (dosing) you or your child needs, or see if the person reacts to a particular food even with the enzymes. Experiment with adjusting doses and food amounts for tolerance as well. Then proceed in the same way with other foods. Keep in mind whether foods are from the same family and therefore might have a cumulative effect.

Some people have a true allergy to dairy, eggs, or other foods (not a protein intolerance or peptide problem, but a IgE-mediated histamine reaction) and are not able to reintroduce these even with enzymes. Allergies can be quite serious.

Peanuts, seeds (such as pumpkin, sunflower, and poppy) and other nuts continue to be problematic for some people to reintroduce even with enzymes (raw seeds may contain enzyme inhibitors). One mom needed to continue eliminating orange juice after starting enzymes. Another person was able to have casein and gluten, but found some corn products still problematic.

A person may still have problems with certain chemicals such as sulfites, dyes, or nitrates. Corn syrup can contain sulfite residues, so you may need to eliminate this as well as other sulfites. Sometimes the quantities need to be limited. Phenols, including artificial additives, are a common problem.

One mom talked about eliminating phenolic foods and food additives with her son. She added No-Fenol when it became available and her son did quite well with it. Thinking the progress was because of the fiber digestion and bacteria control, she slowly slacked off on eliminating the food additives she had been avoiding previously. Over time, her son just didn't seem to be doing as well. A bit jumpy, difficulty getting to sleep again, unusually forgetful, dark eye circles, etc. Then it occurred to her the enzymes were likely helping quite well with actual food content and not the non-food additives. The build-up effect was sneaking through. So, she ditched the artificial stuff and additives again, and her son did much better once more.

A food or eating problem may also be entirely a sensory issue. Watching to see if there are patterns to rejected food may be helpful.

In the end, each parent is the best judge of how their child is responding, and what is helping their own child or themselves. Also, enzymes can be used with any type of diet, so it is not an all-or-nothing situation. There are many combinations of sometimes-enzymes and sometimes-food eliminations. It is up to the individual to decide how much food elimination they want to do, and to what extent they would like to use digestive enzymes, if at all.

No matter how good your diet is or how many supplements you take, if what you do consume never gets digested, absorbed, and used by the body, it won't do any good for your, or your pet's health. It is also interesting that many of these special diets are work-arounds for a lack of specific enzymes of one type or another, or insufficient food breakdown. Thinking of all the digestive ailments that Pottinger's cats developed when eating a diet lacking in enzymes seems related to many current illnesses and behaviors being treated with special diets. No wonder supplying these enzymes contributes to such improvements; sometimes reducing the need for a special diet, or eliminating it altogether.

Hypoglycemia

Several individuals find they or their children are susceptible to hypoglycemia, which means low blood sugar. Hypoglycemia is an often overlooked situation that can be improved with some simple attention to the diet. This situation can sneak in at times because of all the eating issues and digestive problems people with autism and neurological conditions have. Hypoglycemia frequently accompanies *—Alex* yeast or bacteria overgrowth.

Glucose is a type of sugar found in the blood, and is the food for the brain. When we eat and digest food, carbohydrates are broken down into glucose, which is absorbed into the bloodstream. The bloodstream carries the glucose to every cell in the body. Unused glucose is stored in the liver as glycogen. As the amounts of glucose in the bloodstream rise after eating, the pancreas is signaled to release a hormone called insulin (as well as digestive enzymes). The insulin causes the body cells to absorb the glucose out of the bloodstream, which drops the level of sugar in the blood. If too much insulin is produced, the amount of glucose leaving the blood is greater than the amount coming into the bloodstream. The net effect is the blood sugar level becomes too low. If this level of low blood sugar persists, then the brain does not get the food it needs and the other parts of the body do not get the energy they need. The brain depends on glucose as its only source of fuel. The body adjusts to constant low blood sugar eventually by breaking down muscle protein to feed glucose to brain cells, and breaking down fat to fuel the other cells in the body.

When blood glucose is low, it needs to be replenished with a meal or a snack. The person may just feel hungry, or experience an intense sugar craving triggered by the body's attempt to raise the brain's glucose level. If you eat a source of 'quick energy,' such as a candy bar, a cola beverage, or other 'junk' food, your blood glucose concentration may rise too high too fast. The pancreas overreacts and secretes too much insulin, which has the net effect of pushing even more blood sugar out of the blood. Eating a complex carbohydrate food (cracker, bagel, cereal, fruit, vegetable, starch or grain product) is recommended. This reduces the more intense swings in insulin production. Adding protein or fat along with the carbohydrate is even a better choice. Protein or fat slows down the digestion of the carbohydrate. Something as simple as giving a little protein along with all snacks and meals can reduce hyperness and moodiness in sensitive children.

Eating smaller meals more frequently (about every three hours) helps maintain a consistent level of blood sugar. Avoiding sugar and sugar-containing foods and beverages may help, especially on an empty stomach. Try to eat at the same time every day and avoid skipping meals. People with very sensitive nervous systems may react to smaller changes in blood sugar level. My neurologist made a point at our very first meeting for us to always try to eat at the same time, go to sleep at the same, and get up at the same time every day because it helps calm and stabilize the nervous system.

During episodes of low blood sugar, the individual may have symptoms such as anxiety, hunger, dizziness, irritability, weakness, shaking muscles, concentration problems, moodiness, hyperactivity, vomiting, and racing heart. There may be marked personality changes, and they may even seem intoxicated. You can also expect any of the classic symptoms of depression. Symptoms usually appear two to five hours after eating foods high in glucose or carbohydrates. Usually, eating quickly with a properly balanced meal or snack can relieve these symptoms.

In diabetes, the pancreas either does not secrete insulin, or not enough insulin, or the body does not use the insulin produced effectively. The glucose sugar levels in the blood remain high because there is not sufficient insulin to move the glucose out of the blood and into the cells to be used. Some people need to give themselves insulin to assist this process. If the amount of insulin is not regulated correctly, too much glucose can be removed from the blood and result in hypoglycemia. This is why a person with diabetes must constantly monitor their diet, food intake levels, and blood sugar levels throughout each day.

A diet or lifestyle high in refined sugars, highly processed foods, caffeine, emotional stress, or a combination of these factors can cause reactive hypoglycemia. Refining foods strips them of the necessary nutrients and enzymes for their digestion and metabolism. These things stimulate the production of insulin without contributing to nutrition for the body. The net effect is that blood sugar levels are generally kept lower than normal (except right after eating). Eventually any sugar from any source will trigger the pancreas to secrete excessive amounts of insulin. This is how some forms of diabetes develop over time.

If you think hypoglycemia is a factor, you may want to look at a glycemic index. This is a list showing how certain carbohydrates affect blood glucose. A higher glycemic index number indicates a greater rise in blood sugar, and a lower number means a slower rise. This may help in making food choices. Other factors involved in hypoglycemia, such as the endocrine system and adrenal function, may be equally important and warrant investigation if hypoglycemia is a serious concern. Hypoglycemia may be lessened or alleviated with the removal of heavy metals and pathogens, or as proper gastrointestinal function is restored.

A related issue is when a person is intolerant of sugar. Research indicates a fair number of people with neurological issues and digestive problems may have a deficiency in the digestive enzymes that break down sugars (of which there are several types). A sugar intolerance is common enough among people without autism as well, due to the high sugar content in the average diet. Many processed foods contain extra sugar and we may simply be taking in more sugar than we can adequately process out at a time. Taking in excess sugar can have profound effects on metabolism, neurology, and behavior. Research shows that taking more amylase enzymes helps reduce the sugar load in the bloodstream (Mohan, Poongothai, and Pitchumoni 1998).

The seven-month mark came

No crashing. Not one.

A few times, someone had been using enzymes with good improvement for some time and then experienced some 'regression' or adverse behavior. The cause was always discovered to be some other factor totally unrelated to the enzymes. Identified causes included family changes, unrelated illness, new behavior therapy started, other supplements, and change in the weather. If the enzymes did not work out, that was usually apparent from the very beginning.

One mom began regular use of enzymes and saw amazing improvement very similar to the excellent improvement reported by others. After a couple months, she then began an environmental toxin removal program. During this time, her son regressed greatly (that was alarming!). She then discovered a yeast overgrowth problem had emerged. She worked to get the yeast situation under control, all the while continuing with enzymes. When the yeast overgrowth was successfully treated, her son once again regained all of his previous progress plus showed even more improvement.

This illustrates the point that if you are using enzymes, start some other therapy, and see the gains from enzymes 'disappear,' it is very likely it is just a temporary result of the other issue. The improvements tend to 'reappear' as you continue with or correct the other issues, often with much greater gains than before.

A few parents seeing only low or moderate improvement when starting enzymes also saw greater improvement when they proceeded on to a detoxification program or tackled a severe pathogen problem. Remember that enzymes are only a part, perhaps a wonderfully powerful part, of an overall treatment program. Enzymes tend to remain catalysts still, greatly speeding up the healing process, making the entire program much more efficient.

At this point, there were well over one thousand members in the enzyme group and many hundreds more people using Peptizyde and Zyme Prime with various neurological or gastro-intestinal concerns. Other enzyme products were being explored now too. Many folks had been on enzymes for months doing better than ever. The Great Regression never arrived. So ended the waiting game.

Cindy and I seem to have very complementary personalities and abilities. The hand of Divine Intervention must have been at work bringing us together. Many times her emails brought just the laugh I needed, or the guidance, or encouragement, or the insight. Her experience with people and special education nicely balanced my interests in the technical and science side of things. She is the best!

The seven-month mark came and went without much fanfare. By now, it was fairly well established that if you saw good results past the first three to four weeks, they would 'stick,' that is, the positive effects continued consistently. If anything, most people just continued to see ever-increasing improvements as they continued to give enzymes. This was not a fluke or a passing fad. It was not a mysterious leap of faith. It was solid, old-fashion health restoration.

I wrote up a new summary report containing 109 new cases and other findings since the last summary at four months. Did the trend hold up? How did the numbers look over time?

Percentages (4–7 months)
100%   109 Total number of new individuals using enzymes
 92%   100 Number of those with positive results
  5%   6 Number of those with negative or no results
  3%   3 Number of those with inconclusive results

Total percentages (0–7 months)
100%   260 Total number of individuals
 90%   235 Number of those with positive results
  6%   14 Number of those with negative or no results
  4%   11 Number of those with inconclusive results

Looking *real* good! Additional results are in Appendix A, Enzymes in Action – Seven Months of Real-life Use. The success rate of the 109 new cases was now 92 percent, which put the overall average at 90 percent. This reflects different degrees of improvement in different areas as described by the individual family. This increase was basically due to people getting more information about enzymes, and more was just known overall about how to implement enzyme therapy and what to expect. Descriptions of and reasons for the negative and inconsistent results are given on pages 336-337. In the end, about one in ten people trying these enzymes did not come away with favorable results.

Enzyme therapy for autism and related conditions now has a solid place in biological alternatives. We have well-researched explanations for our observations, and a nice set of guidelines to help people get started and obtain best results. Change continues.

A continuing goal is to look for particular formulations or enzyme combinations that will help specific subgroups, if possible – such as for viruses, identified metabolic dysfunctions, seizures, etc. There might be particular enzyme formulations for those with particular sensory issues, or those with language issues, etc.

The enzyme forum has really become the Enzymes-and-other-Alternatives-Applied-in-a-Practical-Way group. There is a constant influx of positive information, progress, and support…and enough of the other kind of moments to keep things colorful. Some of the most popular articles written up are the Guide to Comparing and Buying Enzymes and Mixing Suggestions included in this book, and one for giving enzymes at school.

Enzymes at school turned out to be a challenge all its own. It became the main challenge seen by parents of 'older' school-age children. Despite all of the anti-enzyme tensions, all the harassment and hurdles, and all the attempts at ending this alternative we had struggled through so far, this was one of the few times, I became absolutely…frustrated? No, that's not the right word. Furious? No, that's not it either. Livid.

Yes, I was livid!

CHAPTER 20

Enzymes at School
– Teaching the Fundamentals

My boys were in first and third grade at the time we started enzymes. They blossomed to life, we were ecstatically happy, their Taekwondo instructors were very impressed, and the local elementary school would have none of it. Huh?! Whoa! Back up.

The boys were doing well. We were thrilled. I sent the boys to school. Their grades and disposition improved. I sent the very safe, effective, lovely little enzymes to school for them to eat at lunchtime, and I was promptly notified of school policy on the matter. Many schools require permission from parents or a doctor's note before they will give any over-the-counter supplement or medicine in a school setting. Digestive enzymes usually fall under the same policy as giving aspirin or vitamins. This is for safety and liability reasons. For one, it protects other children so they will not be picking up fallen capsules or tablets and swallowing them. Second, the school wants to have a complete record this is what the child is supposed to be taking on request of a caregiver, and that, in fact, the child did take it if required.

Our current school nurse carefully records the time of day and the exact dose each and every time she gives supplements or medication, even if it is something the child takes every day throughout the entire school year. If either boy does not arrive in the nurse's office on time, she will try to hunt them down (sometimes the class has a special project and they get delayed). If they take the enzymes much later than scheduled, she calls me each time. Further, the school does not want to be accused of making medicinal decisions on their own without authorization, or based just on the word of a young child.

My sons were not allowed to carry the enzymes in their pockets and take them at lunchtime themselves. I tried this a few times in the very beginning but Fidgety-Older-Son was twiddling with them in his pocket and dropped the capsule a couple times on the floor, which is exactly what the school does not want for very valid reasons. Fine, so the school nurse can give them. She said I needed to have a doctor's prescription for her to give them. Fine. I called the doctor's office. They said they could not write a prescription because enzymes were a safe, over-the-counter food and therefore not a drug for which they could issue a prescription. I explained this to the nurse. The nurse repeated her position. I called the doctor's office again. They repeated their position. I was livid. My son's teacher said teachers were not allowed to give anything, only the nurse. The nurse said I could come to school every day and personally give the boys the enzymes, but that was the only other alternative.

Now, I am aware that many other parents have spent years mastering the intricacies of negotiating with school systems, and that I had barely escaped the glorious IEP experience myself, for which I am truly thankful. I have great respect for teachers, the school staff, and all they do. But you would think I was asking them to build a private classroom onto the school building and provide individual tutoring! I was saving them some bucks after all. I could insist they provide my children a casein-free and whatever-else-free meal everyday. Something the boys could eat without getting ill. I could insist they provide special aides to help the boys should they regress wildly by not having enzymes and medications. Wasn't that where we were before? I did not think it was asking that much. After all, medications were routinely handed out every day.

This all occurred at the beginning of our experience with enzymes and, mercifully, toward the end of one school year. I had all summer to consider new options, devise new tactics. I was planning my next move. And then another very remarkable thing happened for us.

That nurse left for another school! Well, this was an interesting turn of events. Now I had a fresh chance at working with the new nurse. I also took notes from Cindy who had been very successful in getting her school to cooperate in giving her son enzymes. This time when I went to talk with the new nurse things went differently. Each school will have different requirements so please check with your specific school. Our school required a doctor's note. This time it was explained as a doctor's *note*, not a *prescription*. I found the magic words for getting the doctor's signature were to ask for a 'written order' for the school so my kids could take something that was non-prescription per their parents' request.

Here is a suggestion that may help: sometimes dramatics are effective. This helped me to get the written order that I was not able to get the first time when I just called and asked politely. I followed Cindy's suggestion of printing everything out from the manufacturer's web site. You can also take in a book or two, or print out some of the journal articles given at the end of this book. The idea is to have a volume of information. This is a visual aid as well as being informative.

I went to the doctor's office with almost an inch of printouts, and placed this stack on the receptionist's desk in a manner dramatic enough to ensure it made a lot of noise. Then, not bothering to lower my voice, equally dramatically, politely yet firmly said,

'My boys . . .

['THUNK!' planting a bottle of Peptizyde down on the desk]

. . . need these enzymes . . .

['THUMP!' down with a bottle of Zyme Prime]

. . . in order to eat food at school without getting sick all over the place!

[this draws a visual picture most people can relate to – you can add in other details to enhance this effect]

This is the research and information on enzymes . . .

[patting the stack of paper firmly, but making sure to whack
the stack with a flat hand so it is very noisy and draws a lot of
attention from as many other people standing around as possible]

I need a written order so the school can do this.'

I left with the written order that day and the rest went very smoothly
after that.

The doctor wrote that my sons needed to take enzymes, one
capsule of each type, 30 minutes before lunch on prescription paper.
This is a written order but it does not necessarily have to be on
official prescription paper. I should mention that my doctor is a fine
fellow who actually was quite understanding. We have a medical
clinic system though. Sometimes bureaucracy gets in the way.

Each day my boys leave their classrooms and go to the nurse's
office 30 minutes before lunchtime and they take their enzymes. The
nurse keeps a bottle of enzymes with their name and phone number
marked on it, so I just send a new bottle to her when one runs out.
Both of my boys are able to swallow the capsules readily. They eat at
school a couple days a week and take their lunches the other days. At
first, I was sending the enzymes in the boys' lunchboxes as I had
previously done (with the capsules tucked in the sandwich or the
enzymes mixed in something). However, it was the nurse's suggestion
to have them come to her office every day to get enzymes whether
they brought or purchased their lunch so everyone would be on a
regular routine and schedule. This has worked out very well for us. In
the beginning, I would also have needed to arrange for them to have
enzymes when party treats were handed out, but at this point they
had 'recovered' to the extent this was not necessary.

As it turned out, our current nurse was very agreeable to learning
about enzymes and finding a workable solution, which brings up
another point. The extent of cooperation you get may depend on the
personalities and attitude of the individuals working with you and
your child. Many people are just not familiar with enzymes and so
do not want to take on the responsibility, or liability, as the case may
be. Medical professionals receive shockingly little education or

preparation in nutrition, supplementation, and any 'alternative' methods outside of traditional physiology and drug therapy. Be prepared to provide a bit of education on the subject as you go. I certainly appreciate that our school is being so diligent about substances on school grounds in general for everyone's protection. I just was not prepared for this hurdle.

Suggestions to help your child take enzymes at school

My sons prefer leaving class and swallowing the capsules. However, this may depend on the child and the circumstances. If your child does not swallow capsules, there are several ideas for sending the enzymes to school in the section on mixing suggestions in Chapter 11. Making the chocolate wafers, frosting balls, or mixing the enzymes in a drink with ice in the morning and sending these to school are favorites.

If a doctor, teacher, therapist, colleague, or other person involved with you or your child is wondering about the scientific validity and longevity of enzyme therapy, you may want to take them this book and flip to the hundreds of references in the back. Point out there are further books with even more references on how food compounds and digestion are relevant to behavior, mental competence, neurology, health, and ability to function overall. You may want to print out a few journal articles and get some of the other books as well.

If you still meet with resistance, here is another option. Dana has been instrumental in the enzyme group as well as with many other groups. She is quite familiar with legal matters and has written the following information and ideas for parents should it be needed. This is not meant in any way to confer legal or medical advice, just some guidelines to consider and intended to be helpful. Dana writes:

> If your child attends a public school under an Individual Education Plan (IEP), or a private school as a public placement under an IEP, then the best thing is to include any diet, supplement, enzyme, and medication issues in the IEP. Most parents of younger children use the words 'parent provided food only' for diet issues. If language addressing your child's specific situation is included in the IEP, the school is required to follow it. If it is not in the IEP,

the school may or may not be accommodating of your desires in this area, and even if the school is accommodating, it is not required to do so.

My suggestion regarding the use of enzymes is that I think the best way to give them to your child during school hours is to freeze them into ice cubes and then put those cubes into a juice container, or dissolve the enzymes right into the juice and then just partially freeze the juice itself. This avoids all the hassles of the school nurse, etc., and your child is not singled out by his friends for any special requirements like going to the school nurse before lunch, and does not have his instructional time interrupted. If you would like your child to take the enzymes as capsules, and your school has a non-prescription medication form, then request that form and complete it for the use of the enzymes you use. You need to do this even if your child has an IEP. Because most school nurses will not be familiar with enzymes or their uses, it would be wise to attach to this form copies of the supplier's ingredient lists, usage information, and instructions for consumption. This can be printed off from the company web site, or other printed material you may receive from them. You would provide the bottles of enzymes to the nurse in their original containers, along with the forms and attachments indicating why your child needs them, when he or she needs to take them, how much to take, and how to take them.

Remember that it appears to take some enzyme capsules approximately 30 minutes to dissolve (veggie capsules), so your child would have to go to the nurse approximately 30 minutes before lunch, to take the enzymes. This, in my opinion, is not the best option, because it singles out your child for 'different' treatment, and other children may tease your child for special treatment, either good or bad. Plus, it does disrupt his class hour immediately before lunch, making it likely that he will have difficulty learning the material which is interrupted. This is also not a good idea if your child has transition difficulties, to leave the classroom, return, and then leave again.

You can request that the nurse perform and/or assist your child to open the capsules and mix the contents with something to take immediately before lunch begins. This to me is preferable to leaving class early, then returning, and then going to lunch as usual. However, it does pose its own problems, primarily that many nurses may not consider 'tampering' with the capsules as something desirable or within the scope of their licenses. A school cannot object to you wanting your own children to consume something that is not illegal, but the school will be concerned about the idea that another child might accidentally or intentionally consume some also. However, if your school is requiring a prescription or other physician approval for enzymes, which are a non-prescription item, and the non-prescription medication form is not acceptable to your school for whatever reason, first ask for that policy in writing. Then you can use a letter similar to the following one, if for some reason you are unable to obtain a physician notice, or need another option.

Sample letter:

[Name], as parent/legal guardian of [child's name] require my child to consume [name of enzyme or whatever] at the beginning of lunchtime [or whenever you specify]. This [name of enzyme or whatever] is a non-prescription legal food supplement which is approved for distribution in the United States by the FDA. It is not a medication so does not fall under the school's medication policy as outlined to me in the attached [attach that policy statement]. It would be considered a 'nutritional supplement' per the attached policy statement.

My doctor has written a prescription (or written order) that my child is required to take this product during school hours.

I am attaching information from the manufacturer/distributor [name of manufacturer/distributor] on [name of product] which includes (1) the nature and purpose of why I am requiring my child/ren to consume this product, (2) a statement of the 'dosage' commendations/requirements of this product specific to the requirements ofmy children, and (3) a statement that this product has been tested and it is not known to have any negative effect in the unlikely possibility that another child accidentally ingests this product.

I believe school policy would be to have this product [name of enzyme] provided to my child/ren through the school nurse, but if this is not a correct understanding of the school policy, then it is my decision that I shall be providing this product to my children in their lunches, and they will be responsible for consuming it at the appropriate time. I will provide only a single day's portion of the product, and I will instruct my children that they are not permitted to share it with the other children.

[If your child is not sufficiently responsible to take enzymes himself in any manner, and/or you would like the school nurse to administer the enzymes, then use the following paragraph.]

Boards of education and school district personnel are not licensed to practice medicine, and a school nurse must implement a valid prescription by a licensed physician. My child is under physician prescription for use of this product during school hours for his ongoing medical issues. Therefore, a pharmacy label is not required, and the school policy disallowing the administration of this product to my child by the school nurse, per the attached dosage guidelines and valid prescription, is not in compliance with the law in this state. In order to ensure my child is in compliance with his necessary medical treatment during school hours, as prescribed by his physician, it is required because of his age/level of responsibility, that the school nurse be responsible for providing this product to my child. This letter is to formally inform you that I am requesting an emergency IEP meeting to include the provision of this product to my child by the school nurse, at the times indicated above.

[If your child is sufficiently responsible to take enzymes himself, or if you will be mixing into juice or food for him to eat, use the following paragraph.]

Because the school policy indicates 'Health Food products and nutritional supplements and/or treatments will not be administered by the school nurse or principal's designee,' it is my decision that I shall be providing this product to my child in his lunch, and he will be responsible for consuming it at the appropriate time. I will provide only a single day's portion of the product, and I will instruct my child that he is not permitted to share it with the other children.

[Use the following statement if it is true for your situation.]

However, I would prefer that the school accommodate my child's medical requirements by modifying the school policy to allow administration of this product to my child by the school nurse.

[Then, end with a statement like this.]

Please let me know the school policy on accommodating the individual requirements ofmychild regarding our doctor's prescription for use of this product at school, whether the school nurse shall be responsible for providing this product to my child, or whether my child shall be responsible himself for consuming it. If I do not hear back from you by [date], I will assume that school policy is *not* to have this product provided to my child through the school nurse, and I will take the action as I have described above.

————- *end of letter*

If you find it necessary to provide the enzymes to your children and *not* through the school nurse, give copies of all this correspondence to the school secretary and require that it be included in your child's file. This is important for zero-tolerance policies. If your child is found on campus with something which looks like drugs, you will have a lot of explaining to do. Your child's file can verify what they are taking, and although you will have a fight on your hands, your child should not be required to accept an involuntary transfer (for the more draconian zero-tolerance policies). This does not apply to private schools and daycare facilities. You will need to work it out with any private providers you use.

A few additional comments: First, instruct your children not to call the enzymes by 'cute' names like 'peps' or 'zymes' or something like that. You will find yourself fighting a battle over your school's zero-tolerance drug policy. You will probably not lose the battle, but it is not a battle you even want to fight, because your child will at least receive a one-day suspension, if not more, before you can get it straightened out.

Second, if your child will have snack times during his or her school day, you will have to consider those times for enzymes also, or provide appropriate snacks for any diet you are on, etc., and include this information in the IEP and/or enzyme instructions as indicated above, in addition to the lunch issues.

Third, if your child is young and/or in therapy classes like Applied Behavior Analysis (ABA), where foods are used as reinforcers, you want to include the types of foods in the IEP. I do not believe the use of enzymes would be appropriate for this sort of situation because they may last more than an hour and/or be performed throughout several hours of a day, and enzymes are not effective beyond an hour or so. But if you want to use enzymes before therapy sessions, that should also be included in the IEP, or give the enzymes yourself if the therapies are the first part of your child's school session. I would personally try to move away from foods as reinforcers, and find non-food items for this purpose. Here is a link with ideas for reinforcers: *www.danasview.net/reinforc.htm*

Fourth, if your child is entitled to a free or reduced-cost school lunch, then he or she is also entitled to a free or reduced-cost lunch that he is actually able to eat. To enforce this requirement on an unwilling school [if your school will not agree to include a special dietary school-provided lunch in the IEP], you will need something in writing from a medical professional which indicates the dietary requirements of your child. Then use that writing to include your child's food restrictions in his IEP, and also include the language 'school-provided' if he or she is so entitled.

Dana has a wonderful web site with a great wealth of information and many, many links that are helpful with many aspects of special needs: *www.danasview.net*

School evaluations

A school may be more receptive to learning about enzymes if they can see this therapy is contributing to the child's performance. Although this is beneficial for any therapy, it may be more important to do this with enzymes simply because of the lack of education in this area. If a school or therapist (or a relative) really balks at giving enzymes, suggest a trial period as a 'test' where everyone will review any progress at the end. A three-month trial would be good, but a shorter period will work if that is all you can negotiate. Fortunately, most people see noticeable results by the third week.

Cindy Kelley is a former special education instructor. She created and took the following questions with her to her son's teacher conference. They identified a list of weaknesses her son showed before starting enzymes and then reviewed it later after using enzymes. You might want to adopt a similar list of questions when discussing evaluations of your child with others. This illustrates how a therapist evaluates. Keeping a record of these over the months or years will help you, and others, better see how your child is improving. Cindy writes:

Below are my notes from my son's teacher conference in the third grade. It documents all the changes reported by my son's classroom teacher and speech therapist. I am not comfortable telling people my son is 'recovered' by the enzymes, but since starting enzymes, he would no longer qualify for the diagnostic criteria for Asperger's Syndrome. My son improved in all areas except stimming, although one could say he did improve in stimming also, because he no longer does it at school, only at home. His stims include loud repetitive noises, so they are distracting to the family. Otherwise, I would not be concerned about them. Stimming alone would no longer qualify him for the Asperger's label. His other areas of eye contact; facial expression; peer relationships; sharing of thoughts; social/emotional reciprocity; preoccupation with restricted areas of interest; and inflexible adherence to specific, nonfunctional routines/rituals are no longer considered impaired by his teachers, and have improved to a degree where my son appears 'typical.' Some areas may have weaknesses, but not impairments, and one could argue that we all have weaker and stronger areas, which give us each a unique personality.

November 7, 2001 Parent/Teacher Conference
Tonight was my son's parent/teacher conference and IEP for speech/language services. One year ago, he was evaluated by the school district. He is now eight. At the time of last year's evaluation, he was casein-free, gluten-free, soy-free, corn-free and off many other foods. At that time, he had already been on the diet for over six months and afterwards continued another six months for a total of a year. We saw many significant gains from the diet. He has been taking Peptizyde and Zyme Prime enzymes from since

late April and has now been off the diet with enzymes for over six months.

To compare his progress since his evaluation a year ago and to try to objectively assess the effectiveness of the enzymes, I went through his testing and wrote down each weakness. I asked the classroom teacher and speech therapist for a yes or no on these previously identified weaknesses. Keep in mind that each of these is listed because it was a deficit assessed through testing before enzymes. A few of these are repetitive since this was compiled from numerous assessment tools. You can modify this for your own situation.

My Son, age 8, Second Grade, November 7, 2001

Previous Weaknesses Observed in Student with Teachers' Current Observation:

1. Stares off or appears to look through people for prolonged periods: No longer observed.

2. Fascinated with screen savers on monitors: No longer observed.

3. Frequently scrutinizes the visual details of objects and can become distracted by visual stimuli in the environment: No longer observed.

4. Frequently lacks a startle response and does not usually localize toward the sound: No longer observed.

5. Does not always respond to his name: No longer observed (not a problem).

6. Hyper and hypo responses to sounds: Much better.

7. Easily distracted by auditory stimuli: No longer observed (not a problem).

8. Twirling, hand flapping: No longer observed.

9. High tolerance to pain: No longer observed.

10. Difficulty with gross motor such as kicking, throwing: Yes, some.

11. Writing is difficult: No concern.

12. Often does not respond to gestures for hi and goodbye: Sometimes.

13. Voice lacks appropriate inflection and is high in pitch: No concern, much better.

14. Sometimes displays slow processing in that he may respond to a question after other questions have been presented: No longer observed.

15. May continue an activity long after it is no longer appropriate: No longer observed.

16. Difficulty with abstract learning: A little bit – some difficulty with math logic such as, 'Jack makes $2.75 and hour and works for two hours. How much did Jack make?' This was the teacher's example. Usually no concern, though.

17. Generalizing skills and making associations appear difficult for him: No longer observed.

18. At times may struggle with concept retrieval: No longer observed. Understands well.

19. Does not typically respond to the facial expressions of others: No concern, 'I think he would understand the expressions,' classroom teacher.

20. Has difficulty with appropriate play skills and social interactions with peers: No longer observed. Seeks out children to play with.

21. Lacks friendships: Not true, he has friends.

22. May engage in ritualistic behaviors, such as twirling: No longer observed.

23. Very attached to the computer: No longer observed.

24. Would prefer to be involved with inanimate objects, such as the computer: No longer observed.

25. No social smile: Not a lot of smiling, but seems content.

26. Avoids eye contact: No longer observed, not a concern.

27. May learn a task but forgets it quickly: No longer observed.

28. Does not follow simple commands: No longer a problem, follows commands.

29. Has poor visual discrimination skills: No longer a concern, does well.

30. Is not responsive to others' facial expressions: No longer a concern, does ok.

31. Lacks a startle response: Do not know.

32. Often appears anxious: No longer a concern.

33. Seems unaware of dangerous situations: Do not know.

34. Stares into space for prolonged periods: A little, a tiny bit.

35. Eye contact: Could be better, but not a problem, has improved significantly.

36. Initiating interactions with peers: Not often, but some.

37. Initiating interactions with adults: Yes – observed.

38. Distracted by computer monitor: No.

39. Sit properly at desk (during an observation, sat with legs up in chair): Now sits appropriately.

40. Voice affect: Improved significantly. Rate is a little slow.

41. Handwriting: A weakness, but not a big concern.

42. Walking in line: No longer a problem.

43. Following directions: No longer a concern.

44. Following routine: No longer a concern.

45. Eating lunch: No longer a concern.

46. Sedentary: No longer a concern.

47. Tolerating change: Does fine, no longer a concern.

48. Mental math: Not a concern, except some difficulty with abstract math as mentioned above.

49. Sentence structure: Perfect, no longer a concern.

50. Pragmatic language: Only concern now in understanding intentions of others, but not a big problem.

51. Word structure: No longer a concern.

52. Speaking abruptly: No longer a concern, 'That is just gone since the enzymes.'

53. Articulation /f/ for /th/, /sh/, /ch/, /s/, /j/ distortions: Occasionally missed in spontaneous conversation.

54. Slow rate of speech: A little slow.

55. High pitch: No longer a concern.

56. Reading unspoken verbal cues/body language: Appears to understand, no longer a concern.

57. Expressing personal thoughts and feelings: Could do more, keeps feelings on the back burner.

58. Understanding of how his interactions are perceived by others: No concern.

59. Social relationships within the school setting are not generally within age appropriate expectations: No longer a concern.

60. Difficulty when required to sit and listen, may react passively when asked to complete a familiar task: No longer a concern.

The teacher shared that over the summer she thought to herself, 'Oh, I'm going to have the Kelley's son…,' with a loud sigh like when you feel hopeless. She had heard about the difficulties. 'But,' she added immediately, 'the progress is unbelievable! The problems are just not there this year.' She told us our son is religious about taking the enzymes and would not dare eat even the tiniest treat without them. This is a huge contrast to what we went through with his resistance to the restrictive diet.

The teacher also shared that his reading skills and more importantly comprehension are excellent. She said his critical thinking skills on the current reading test were also excellent. The speech therapist also commented on the improved critical thinking skills and referred to the reading test as evidence. She had some concern about eye contact, but said that she was not sure how much of a problem it was because he appears not to be paying attention at times, but actually is and can answer all questions. She said she thinks he is not always looking at her because he is always trying to do two things at once and can, which very few other children can do. Although he still stims at home at times, the speech therapist said she said she no longer sees him stim at school.

The speech therapist said she worked on my son's high-pitched voice last spring using a computer program. However, he did not progress until starting the enzymes. She said she believed the enzymes were responsible for the improvement because it was immediate. She also shared that he can now follow directions given the first time and directly attributed this to the enzymes. She said the difference is 'incredible.' Before enzymes she used to say everything three to four times and then have to touch him and maybe say it again and then perhaps he would follow through. Now she says directions only once.

'Incredible,' she said, at least two more times.

CHAPTER 21

Little Princes at Play
Once Again

One of the most important things I hope to convey is that you as a parent and an individual are the best person to evaluate your child's or your own treatment program and progress. Because many of the aspects of neurological conditions vary so greatly and have multiple causes, only you will be able to determine ultimately the effectiveness of a given treatment program. There is no one else. This is not a reflection on any person or profession; it is just the nature of dealing with a vague, elusive spectrum of conditions.

Amber was continuing to follow her instincts on which therapies to pursue while evaluating her son's reactions to each one when she came across the enzyme alternative. Amber has an interesting story because she received direct feedback from a very reputable clinic on the effectiveness of her 'Mother's Protocol.'

> My little guy, Spencer, was 'diagnosed' with autism at 17 months, although the doctor said she hesitated to label him at such a young age. She did say that he was lucky we intervened at such an early age, as early intervention was a key component to future success. My son has never been immunized, and we suspect that antibiotics and ear infections triggered his condition. We immersed ourselves in research about every

therapy and treatment that traditional and non-traditional medicine recognizes.

Spencer started occupational and speech therapy right away. Several months later, I found out about the new types of enzymes through research and started my son on those without ever trying the casein-free, gluten-free diet. I intended to try this diet, but, as most of us know, it becomes a food choice that defines your entire lifestyle. My biggest concern, was that there are no long term studies that tell us how kids are doing after having these major food groups removed from their developing bodies forever. It seemed to me that I was hearing about too many other food issues coming up after putting kids on this diet, and my feeling was that the diet might actually be causing these issues – not just uncovering issues that were already there.

So, I ordered Peptizyde and Zyme Prime enzymes and began giving them as directed. What was odd about the experience is that I didn't think we saw any progress. I was looking in the wrong direction. I was expecting my son, 2 months shy of 2 years old, to begin talking and jumping and increasing his skills at a phenomenal rate, as I had heard from others can happen with these enzymes. I heard that one bottle of each would clearly tell us if Spencer was going to benefit from this product. At the end of the first bottle when we ran out, I decided that we had not really seen any change in Spencer, so we would stop using the enzymes. I remember this so clearly because it was a Thursday afternoon. Well, the next day I noticed that Spencer was just being pretty difficult to get along with. By Saturday morning, he was a child I remembered from my recent past, but who had gone away, gradually, over the month that we were using the enzymes. He was a terrible beast and I wondered how we ever lived with him before. I immediately went to the health food store to buy more enzymes, the best ones the store sold...after all, they were JUST enzymes, weren't they?! Spencer was not responding with these.

We lived with Spencer's awful dispostion for an entire week from the time we ran out and received more of the special new ones. We saw amazing improvement right away upon reintroducing the enzymes. I always give my kids good, hearty breakfasts to get them going. So, Spencer ate oatmeal every morning. Warning: don't ever, ever mix proteases in your child's oatmeal. You will have nothing but oatmeal soup in 5 seconds flat. Instant food breakdown. Think I am exagerrating? Try it.

Later on, once again our next shipment of enzymes was delayed due to events beyond my control. This time Spencer did *not* have a negative reaction to the enzyme withdrawal. I thought this was very odd and continued to observe him very closely. Then, once our shipment arrived and we re-introduced the enzymes, he did *not* have a positive reaction. This just made no sense. Then one person who also gives lots of oats at her house mentioned to me that oatmeal has gut-healing properties and surmised that perhaps the enzymes and oatmeal combination facilitated intestinal healing in Spencer. It had been two to three months by now, sufficient time for gut healing.

I think the very first *effective* treatment for us were these new enzymes – by far and away. So, now the Happy Child lives at our house again. I found lots of great reasons to love my son all over again. The next *effective* treatment for us was Therapeutic Listening. This is a type of auditory therapy that increases the types of sounds a child seems to respond best to, causing his ears to 're-learn' what they are hearing, and the brain to correctly process what it is being fed. Spencer responded positively from the first day. It was as if the music was helping him to regulate himself and to 'sense' things appropriately. After only four days, he also began using eating utensils primitively. Only the week before, he would throw them at me and scream if I even tried to get him to touch them as I pushed the food into his mouth. On day seven, Spencer began communicating with us. After five months of speech therapy and not a single attempt to communicate in any way, Spencer suddenly began signing 'more' and 'all done.'

There is much more dramatic improvement that I could describe; however, suffice it to say that the real person began rapidly emerging. The most recent effective treatment we tried was cod liver oil following Dr Mary Megson's protocol (www.megson.com). At 2 years and four months, Spencer's vocabulary jumped from only 3 unintelligible words to an official count of 175 words, less than three months since beginning the cod liver oil. His cognition has also improved to an amazing degree. He now engages in hilarious imaginative play with his stuffed animals as well as our real cats. He began stringing two words together this week and even answers questions now. No telling to what extent taking enzymes before the auditory and cod liver oil therapy helped with making these other treatments so effective. All we know is that we were not gaining much ground anywhere until Spencer started them, then everything started taking off.

Spencer continues taking several nutritional supplements each day. The regimen we have settled upon is the direct result of the research I have done into the alternative treatments for kids with autism. Spencer also benefits from Epsom salt baths to help a magnesium deficiency.

It is my very strong opinion that these measures have lifted the fog and made Spencer available and willing to benefit from the speech, occupational therapy, and sensory integration therapy that he has received all along, but not improved with…until the implementation of the enzymes. And not just any enzymes, these particular ones.

As I have told every breathing person who would listen to me, I never focused on my toddler with autism. I focused on the fact that he would be a grown person with autism and he needed our efforts for the best quality of life…regardless of the prognosis and regardless of what the 'experts' felt they knew to be his fate. I knew that no one else would care how he ended up as much as I would care. Luckily, I did not have to reinvent the wheel, because there are many fine, committed parents whose experience and wisdom guided me through the chaos of a spectrum condition that has no concrete cause, treatment, protocol, or cure.

Spencer has had several speech therapists, occupational therapists, early intervention programs, and behavioral intervention programs as well. I persisted and would not accept 'no' because he depended on my efforts.…nor would I chase rainbows and become a victim of any unethical opportunists who would prey upon desperate parents who love their children. So, I had to use my critical thinking skills, maximize the available resources, and balance the needs of our son versus the needs of our entire family. Spencer is now just blossoming!

We had our special education evaluation meeting today and the speech therapist said she was astounded at Spencer's progress. She said that parents may hand their child over to a therapist and expect them to 'fix' the child in their one hour or so a week. The better attitude is to recognize a therapist's job is to give parents the tools they need to use during the child's entire lifetime.

I made arrangements for Spencer to be seen by the Pfeiffer Treatment Center in Chicago. This clinic specializes in mental illness, neurological disorders, and chemical imbalances by balancing body and brain chemistry. We received the results from all of the laboratory tests run on Spencer and went in for a consultation. The nurse who ran down Spencer's nutrient levels and indicators of absorption said that obviously we have already been on the right track because Pfeiffer is not suggesting that we add much to what Spencer is currently getting. In fact, they did not even list many of the supplements he is currently getting as even necessary. Maybe it is just wishful thinking, but maybe it is that all of this research and time really is paying off in very tangible ways.

The nurse said that the regimen I have Spencer on is actually better than what they would put him on. They had not heard of Peptizyde, but were very impressed with the results.

This story is important to our family for obvious reasons, but I want to share it with you because:

1. I want you all to see the same wonderful results we have, and to know it can be and is being done.

2. I do not want you to become frustrated and disheartened by the array of supplements or therapies you are 'suppose' to be giving your child. Each person, each situation, each therapy is different. You will eventually find the right program if you do not give up. Trust yourselves as parents, as individuals. Your gut instincts are *exactly* what your kids or you need.

3. For those 'on the fence' about these enzymes, remember that enzymes are providing what your body should be providing anyway. Three weeks for under $50 can produce a lifetime of blessings.

4. Listen to your child's response. Do not rely so heavily on lab analyses and well-meaning specialists that may miss the cues your own child is giving you. Some of the best specialists agree that they themselves rely heavily on the child's particular signals and responses, and this is the preferred way to proceed. The nutritional protocol that Spencer is on, as well as all of the other therapies he is enjoying, were all initiated before we had one single lab test done. You want to treat the person, not just treat the lab results. It is very important to have confidence in yourself and your decisions. There are hundreds of books as well as scientific studies out there, many of which contradict one another. You can easily achieve a major migraine trying to digest the many books, and theories, and treatments, much less trying to implement them all.

I also wanted to give some hope, the one thing that all of us desperately need.

Step Two – What to do about it

Commonly, the first step is trying to figure out what the problem actually is. If you are like me, you will usually want to do this as quickly and decisively as possible because you want to hurry on to Step Two – What to Do About It. Sometimes you can flounder around for years attempting to treat something without an accurate diagnosis. Or you may only have a partially correct diagnosis, like one of the blind men evaluating the elephant. Checking for a yeast, bacteria, virus infection, or other medical condition first may help before giving a diagnosis based just on behavior. There may be other subgroups as well, such as having an auditory processing issue.

However, with neurology and biological systems, we can't always find one distinct thing. It may be the total load. Spectrum conditions are not perfectly defined and time moves on. You may have to set aside the fact you do not yet have a specific diagnosis and a concrete plan of action to implement. Just do what you can, evaluating each alternative in light of your specific situation. Each step may help a little bit more.

I followed the methods that produced positive results for us. If a therapy produced good results for others, and we did not experience the same thing, it does not mean 'I didn't do it right.' Or 'I did not try hard enough.' Or that I should not have even tried it at all. I just need to select wisely, evaluate thoroughly, try it to the best of my ability, and if there are not positive outcomes, move on to something else. (Boy, have I done a lot of 'moving on' in my day!) Some of the things I tried worked for just Matthew, or just Jordan, or just Matthew and me. Several worked for all three of us. Some for none of us.

I also learned that staying flexible was important. Just because we are doing a therapy today does not mean we will be doing it for ever (and probably will not be). Or just because my son needs a medication or therapy today does not mean he will need it the rest of his life. He may, but then again he may not. And either way, this is not a failing on anyone's part. It is something needed at the moment.

There is a balance between accepting the situation realistically, dealing with it today, and maintaining hope for a better, different future. No telling what will pop up. A new medication maybe, or new scientific discoveries. The enzymes we found successful were not formulated until Matthew was nine years old. I try to live each day as best as possible, see and appreciate the blessings of each boy as the individual they are right now; try not to get so wrapped up in the future and miss today, because it will never come again. It is a balancing act, something I achieve better on some days than others.

As you probably know by now, what one person may need to do exactly may be very different from the best course of action for someone else. I have shared much of what helped us, and others with different situations have shared their stories. Some main standouts are listed below.

Enzymes – Even if you do not know exactly what caused, triggered, or contributed to your or your child's condition, enzymes may effectively and efficiently work directly on, or support the restoration of, almost all the identified related biological concerns. Studies and field-testing show they are practical and effective in everyday life as well. The body can heal itself magnificently if given the proper nutrients and materials to work with. Refer to the Guide to Selecting Enzyme Products, Appendix A.

Probiotics – Adding a good probiotic is usually a safe bet. Many companies make fine products. Culturelle is a solid workhorse backed by extensive research. Different people respond better to different probiotics, but some people can not tolerate any probiotic supplements. Yogurt with live cultures is a good source of probiotics in a whole food source which may convey other beneficial nutrients as well.

Medications – There are many excellent beneficial prescription medicines available for different conditions or symptoms. Traditional and non-traditional methods can work quite well together. It is very advisable to look into resolving any sleep problems as soon as possible. Remeber the sleep/wake cylce and managing blood sugar levels.

Treat any pathogens, if needed – yeast, bacteria, parasites. Treating a virus may require special considerations and a specialist. Yeast in particular can substantially impact neurology. Treating pathogens may clear up many other associated problems.

Eliminate remaining problematic foods – Eliminate or reduce any additional foods or chemicals that may not be tolerated even with enzymes. There may be a true allergy. Reducing exposure to problem chemicals as much as possible may reduce the total load on the system even if someone does not show an intolerance for one specific chemical. An inexpensive start is reading labels eliminating additives and basic good cooking with healthy foods.

Detoxification – 1. Eliminate toxins and stressors possibly entering in, and 2. Assist or enhance removal of anything already present that might be hampering healthy Epsom salts go a long way, are inexpensive, and easy to do.

Other nutritional supplements – Caution and a very, very conservative approach are usually best. Thoroughly research what you need for your particular situation. Brainchild Nutritionals makes an exceptional multivitamin/mineral product specifically for autism spectrum and attention deficit conditions which covers a great deal of nutritional ground. Brainchild is designed for those with sulfation issues and the very sensitive person (call for availability). Look for products that are highly absorbable which is a major benefit for those with gastrointestinal problems. Magnesium, zinc, and essential fatty acids may be ones to keep in mind. Many good sources for nutritional supplements are available. Whole food supplements tend to be very beneficial when possible.

Staying in touch – Being involved with some sort of support group and networking with others who are dealing with similar situations can be extremely helpful. You can find a great deal of experience, practical suggestions, ideas, advice, and support you may not find in a waiting room.

Behavioral therapy – As you are pursuing the biological aspects, also consider and pursue any appropriate behavioral therapies. Sensory integration, floortime, auditory therapy, exercise, and others may be very beneficial. Behavioral and physiological therapies are two sides of the same coin working together. It is like making sure your child gets a good breakfast and does not have a cold before school so they will be at peak learning ability. But eating the best breakfast in the world isn't going to put math or reading knowledge in your head.

All of this is an overall balancing act with no one right way to go. The double-sided approach is to reduce the size and weight of your 'elephant' while strengthening your 'canary.' Stay flexible, monitor the situation constantly, re-evaluate regularly, and adjust as necessary. If the immune system gets too stressed, it leaves an opening for other things to develop or get out of hand. And you will not be gaining ground overall. It is a very subjective judgement call to decide what is 'too much' or 'too unwell' or 'too much bacteria still' to continue. Incorporating all of this into your daily life is just a further challenge. There is a great deal to be said for quality of life as you go, too.

You probably noticed a distinct lack of discussion on laboratory testing. There are so very many tests it would require an entire set of books to cover. Some practitioners and commercial labs are pretty gung-ho insisting on running an entire spreadsheet of tests before you begin (and maybe pay up front, not covered by insurance). Although this method generates a lot of numbers, many families, as well as doctors, find that this may not prove terribly useful for improving the patient's health. Many of the tests are not conclusive in giving you accurate direction for a treatment either - false positives, false negatives, two different results. The tests available just may not be capable of accurately measuring what you are hoping to find out. My view is to wait on testing until I know I need something specific, know what the test will tell me *exactly*, and what it will not tell me exactly, and know what I am going to do with the results once I get them. It is very beneficial to insist the practitioner or vendor provide this information before breaking out the checkbook. This helps everyone stayed focused.

If you start enzymes and a good diet first along with some basic supplements such as a highly absorbable or whole food-based multivitamin/mineral, this may correct many nutritional and metabolic problems that would otherwise show up as deficiencies on tests. Testing is up to each person, of course, and may be necessary some-where along the line. Different practitioners have different ways of going about these things. The idea here is to approach testing cautiously and conservatively instead of running every test possible just because it is there.

Since April 2001, all the previous trends have held steady for over two years at this point (May 2003), plus we have a few more refined guidelines. Families continue to find more manageable and enjoyable ways to assist their children and improve their quality of life. Peptizyde brings to an end the long and elusive search for an enzyme product that could sufficiently break down the potentially problematic gluten, casein, and other proteins. It has proven its worth. This elegant, yet brilliant formulation opened the treasures of enzyme therapy to the world of autism and related conditions. Now interest in enzymes is peaked once again. New enzyme products are emerging, and previous ones are being rediscovered. There is an open invitation to visit the Enzymes and Autism group and see if you find anything useful: *http://groups.yahoo.com/group/enzymesandautism* or see the link at *www.enzymestuff.com*

We are now in the pleasant position of seeing a number of people leave the group because they or their children improved to the extent that they no longer needed this type of support. With more time and experience, we hope to have even better guidelines for using enzymes. Just as drinking lots of water may be core to many diets, enzymes may be fundamental to biological therapy and good health.

The mommy takes a rest

At this point, we have two very sweet, happy, wonderful boys who are doing well in all areas of life, and an equally pleasant mom to match. We have around an 80 percent reduction in our mountain of sensory issues putting them in the 'acceptable' or 'typical' range. That is, sensory issues are more like strong personal preferences instead of debilitating. They no longer disrupt our day-to-day life; quite manageable. I still need my textured socks, and sunglasses on any sunny day. However, I have now found one place I enjoy seeing the color orange – inside a pumpkin pie under billows of whipped cream!

Mike is very, VERY pleased that I no longer have a severe headache every night – the Happy Husband Effect.

Besides the enzymes and medications, I still give a few additional things, such as a probiotic either as a supplement or in yogurt; magnesium as a supplement or in epsom salts; a basic whole food-based multi-vitamin/mineral for general health maintenance; and one son does better with just a little extra zinc. I try to provide everything possible through a healthy diet, including 'good' fats, and reducing artificial ingredients and chemicals along the way.

We still find the floortime method of play/education with its emphasis on more individual attention and exploration best for us. Sensory therapy continues to be a very practical avenue.

Other incredibly beneficial activities included computer programs. The boys spent hours doing interactive educational software programs usually with each other or with a parent. I really think this helped immensely with their socializing, building thinking skills, and conversation. Our favorites included Dorling-Kindersley, Humongous Entertainment, and Living Books children's software which were out at the time.

Watching videos and old movies continues to be a beneficial activity. I realize this is not conventional, but we watch with the boys and explain how the special effects were done, why the characters are probably behaving the way they are, what was going on in the scenes, and how this related to the plot. We used it to teach vocabulary, social interaction, behavior skills, and about life in general.

After being in Taekwondo for three years, we attended one of the promotion ceremonies for advanced belt levels involving several schools which happens once every three months. Matthew was promoting to the level of Red Belt and was named Student of the Month for his level and personally recognized for technique, skill, and personal character. That is a grand achievement for anyone, but we knew the extra load he had carried.

And here is a crucial component: Prayer, in generous, heaping amounts, remains a cornerstone ingredient to our success as well! Thinking about the money we spent that didn't bring any positive results...is something we just need to put behind us now. We even started physiological measures when Matthew was five years old – way past the preschool window of opportunity recommended. I am over 40 now. For me, this is not just a disorder or a condition – it has been my life. So for anyone wanting to know if there is hope for older children...there is!

I think that one of the reasons Matthew showed such dramatic jumps in improvement with both the amitryptiline and enzymes was that the other activities and therapy we had been doing all this time were soaking in as well. Matthew had been listening and learning all along, but was hurting so badly physically he could not express it fully. Sometimes he will now say things like 'Remember when we were at the mall when I was five years old and we were looking at the ferret in the pet store but I had a really bad headache and was all fussed-up? The ferret was really funny when he...' And then he clearly recites the entire adventure there. What I remember was that Matthew was melting down at that time and not paying the least bit of attention to the ferret. But he was. And he remembers it.

He remembers the times I said 'I love you' as he jerked away clutching Yellow Blankie. He remembers the lessons at school though he would not move or talk. He remembers the activities we did with him though he refused to participate. I am very glad I made all these efforts now, because although at the time it appeared he was not getting much out of it, he *was* taking it in, at least some of it. When he finally felt better, all of this came bursting forth.

One of the greatest blessings of all is the simple fact that Matthew is so affectionate now. I get spontaneous hugs and kisses as he passes in the house *all the time*. He says goodnight lovingly and cuddles with his dad, laughing and giggling. He tells brilliant jokes. I watched nine years go by, seeing this child suffer. And now it is at an end.

We have all been evaluated as 'recovered.' My own improvement is just a bonus. Recovered is a very appropriate term in our case. Jordan and Matthew are no longer delayed in any area – physically, socially, academically, cognitively, or functionally. They do well in regular school, even well above average is some areas. They have some close friends and continue to be each other's best buddy. And they are enjoying a very typical childhood.

Are they still unique? Well, of course! Has becoming recovered from the physiological maladies detracted from their special abilities, thinking style, or made them a different person? Not a chance. They are just *more* of who they were before. More happy, more healthy, more creative, more thoughtful, more capable, more talented, more loving, more alive, more in life. No...all that other stuff just hampered them. They have lost none of their special personalities, and have gained all their health. We love them.

Our Little Princes.

Guide to Comparing
and Buying Enzymes

Seven parts to navigating the world of digestive enzyme supplements. This guide outlines the steps and processes to take in reading product labels, matching enzymes to food types, and what to look for in a digestive enzyme product.

Part 1. Why are you considering enzymes?

Select an enzyme product based on the results you want to achieve. Think about the food groups you want the enzymes to break down, and then pick a product that contains the proper enzymes. Products are usually a mixture of enzymes, not just one type. You may need to choose more than one product to cover all the foods you need to break down. Sometimes you may have another goal besides food breakdown. Examples are using a high protease enzyme product to take between meals for inflammation, gut healing, and blood cleansing, or one with a high level of cellulase to help with yeast overgrowth.

Skip right through all the advertising and marketing fluff. Note what end results you want to see and use that to make your decision. You can get some helpful information from a company but be sure to compare this information with other sources as well.

Digestive enzymes are enzymes that break down food into usable material. The major different types of digestive enzymes are:

- amylase – breaks down carbohydrates, starches, and sugars which are prevalent in potatoes, fruits, vegetables, corn, grains, and many others
- lactase – breaks down lactose (milk sugars)
- diastase – digests vegetable starch
- sucrase – digests complex sugars and starches
- maltase – digests disaccharides to monosaccharides (malt sugars)
- invertase – breaks down sucrose (table sugar)
- glucoamylase – breaks down starch to glucose
- alpha-glactosidase – facilitates digestion of beans, legumes, seeds, roots, soy products, and underground stems

- protease – breaks down proteins found in meats, nuts, eggs, and cheese
  - pepsin – breaks down proteins into peptides
  - peptidase – breaks down small peptide proteins to amino acids
  - trypsin – derived from animal pancreas, breaks down proteins
  - alpha-chymotrypsin – an animal-derived enzyme, breaks down proteins
  - bromelain – derived from pineapple, breaks down a broad spectrum of proteins, has anti-inflammatory properties, effective over very wide pH range
  - papain – derived from raw papaya, broad range of substrates and pH, works well breaking down small and large proteins

- lipase – breaks down fats found in most dairy products, nuts, oils, and meat

- cellulase – breaks down cellulose, plant fiber; not found in humans

- other stuff
  - betaine HCL – increases the hydrochloric acid content of the upper digestive system; activates the protein digesting enzyme pepsin in the stomach (does not influence plant- or fungal-derived enzymes)
  - CereCalase™ – a unique cellulase complex from National Enzyme Company that maximizes fiber and cereal digestion and absorption of essential minerals; an exclusive blend of synergistic phytase, hemicellulase, and beta-glucanase
  - endoprotease – cleaves peptide bonds from the interior of peptide chains
  - exoprotease – cleaves off amino acids from the ends of peptide chains

- extract of ox bile – an animal-derived enzyme, stimulates the intestine to move
- fructooligosaccharides (FOS) – helps support the growth of friendly intestinal microbes, also inhibits the growth of harmful species
- L-glutamic acid – activates the protein digesting enzyme pepsin in the stomach
- lysozyme – an animal-derived enzyme, and a component of every lung cell; lysozyme is very important in the control of infections, attacks invading bacteria and viruses
- papayotin – from papaya
- pancreatin – an animal-derived enzyme, breaks down protein and fats
- pancrelipase – an animal-derived enzyme, breaks down protein, fats, and carbohydrates
- pectinase – breaks down the pectin in fruit
- phytase – digests phytic acid, allows minerals such as calcium, zinc, copper, manganese, etc. to be more available by the body, but does not break down any food proteins
- xylanase – breaks down xylan sugars, works well with grains such as corn

Part 2. What types of enzymes are in the product?

Source of enzymes

All digestive enzymes come from two living sources: plants or animals. The plant group includes both those enzymes derived from plant sources (pineapple, papaya, kiwi) and microbial (fungal) sources. In general, plant enzymes are preferable when possible. They offer several advantages over enzymes from animal sources. Plant and microbial enzymes are much more effective in the pH and temperature ranges of the body.

Our pancreas, when working properly, secretes a number of enzymes to digest food as it enters the small intestine. But as we age, or in some disease states, this enzyme secretion may not be adequate to completely digest the food we eat. This can result in pain, cramping, excessive gas, certain food intolerances, and inflammation. Pancreatic enzymes are available by prescription (Creon, Viokase) or over the counter. However, pancreatic enzymes are not stable to the acid conditions found in the stomach, so a good portion of them may be destroyed unless the preparation is treated in such a way, like being enterically coated, so that the enzymes will not be released until they arrive in the small intestine.

Plant and microbial enzymes, however, are stable in acidic conditions. They help digest the cooked and raw foods in the higher pH of the upper part of the stomach, the acidic lower part of the stomach as well as in the alkaline intestines. Digestion in the upper stomach actually mimics the natural process of eating raw foods, which contain some amount of the enzymes needed to break down the food itself. The additional 'pre-digestion' provided by plant and microbial enzymes leaves the pancreas to provide the 'finishing touches' to the digestive process in a less stressful manner. The intestinal tract will be better able to absorb and assimilate the nutrients and vitamins in the meal.

Is an all in one enzyme product better or a speciality product with only a few different types of enzymes?

There are advantages and disadvantages to each strategy and which is 'best' will depend on the individual situation and particular product. Here are some considerations.

'Everything' product – A comprehensive product alleviates the need to think about which enzymes go with which foods, so it may be easier to give just one thing. You can take it for overall digestion whenever you eat. However there are dozens of products claiming to be 'the ultimate' or 'most comprehensive' enzyme product, using very different amounts of different enzymes. Consider any blanket statements like this to be marketing jargon.

Product Toleration – Some people cannot tolerate certain enzymes for a variety of reasons. Having separate formulations allows many more people to enjoy the benefits of enzymes because they can eliminate the enzymes they do not tolerate. Having everything in one capsule makes it an 'all or nothing' deal. If someone reacts negatively to a formulation, and all the enzymes are lumped together, there is no way for the person to fine-tune it, or figure out what is the problem.

Specialized need – It just turned out that because Peptizyde was a very strong protease-only formulation and the other enzymes were all in Zyme Prime, we were able to figure out many very important factors and guidelines for enzyme therapy (such as the celiac issue). You may want particular enzymes just for yeast control. If you are on a diet that contains high fat, you would probably be better off with a specialty product with a much higher level of lipase enzymes than most broad-spectrum products contain (Lypo from Enzymedica is an example). Some people cannot tolerate the fruit-derived enzymes, whereas others specifically want bromelain or papain to help with inflammation. A few have reported taking a a strong protease product to quelch a migraine or cold. Wobenzyme N is all proteases and very popular for immune system support. If you take a complete product while attempting to get the benefits of just one or two types of enzymes, you may have to take much more product which makes this much more expensive. Targeting specific needs may be more efficient and cheaper. Calculate the cost per capsule and per dose for what you need. There is also the basic issue of volume or bulk in enzymes. There is only so much room in a capsule so the all-in-one product may require you to take more capsules just to get adequate amounts of the basic types of enzymes. Some blends have special properties.

Proteases separately – Giving proteases separately has proven to be very beneficial for many people. Since the proteases are doing many other types of healing work in the body, this provides advantages and disadvantages. The advantages include fighting pathogens, eliminating waste and toxins, immune support, etc. We have noticed that negative initial adjustment reactions can be minimized by giving proteases separately and going slower with them, where as this does not seem to be a factor with other enzymes. Many parents have found that being able to dose a strong protease product separately from the other enzymes has made enzyme therapy successful for them. They can use a broad spectrum product lower in proteases first for a couple of weeks to promote gut healing gently and then introduce the stronger proteases. For those with yeast, die-off may be slower and more tolerable. Someone with a severely injured gut can give the proteases slower until the gut is sufficiently healed without having to give up the benefits of other enzymes. This strategy may actually speed up gut healing as well.

Number of enzymes – You will notice that many very good enzyme products do not have every enzyme known to man in them. Having one of everything is not really necessary with enzymes. It may be helpful for some people depending on their physiology and diet, but many people just need ample supplies of the basic ones. Too many different types of proteases may start to cancel each other out. Also, certain combinations of enzymes have synergistic benefits that are not seen if given separately or not in the appropriate combination. This is the 'art' and science of making targeted products.

Part 3. Look closely at the amount of enzyme activity

Enzyme strength is measured in terms of activity. Enzymes may be present, but unless they are functional, they will not do any good. While most food, supplement, and drug comparisons use weight (such as milligrams), the most important measurement with enzymes is the activity and potency of the enzyme. A product label should list enzyme strength in standard activity units rather than by weight. To measure activity of digestive enzymes, tests or assays determine the quantity of digestion that occurs under specific conditions. This activity depends on concentration, quantity, pH, temperature, and substrate.

When you review the labeling on a digestive enzyme package, look for Food Chemical Codex (FCC) units. This labeling certifies that the enzymes went through thorough testing for activity and potency. The American food industry accepts these units as set forth by the National Academy of Sciences. Some companies promoting enzymes list measurements based on dosage, weights such as milligrams (mg), or a other things. Weight, dosage, and any other units do not give any information on enzyme activity – 220 mg per capsule does not tell anything about enzyme activity. You may have 220 mg of nothing, or 10 percent activity or 90 percent activity. FCC labeling is the only national standard for the evaluation of activity and potency of enzymes in the United States. If the product you are interested in only gives weight in milligrams or in units you do not understand, you can call the company and ask about the specific ingredients and activities. FCC Units:

- amylase – DU or SKB (Alpha-amylase Dextrinizing Units)
- bromelain – GDU (Gelatin Digesting Units)
- cellulase – CU (Cellulase Unit)
- glucoamylase – AG or AGU (also AU or AG) (Amylo-Glucosidase Units)
- invertase – IAU or INVU (Invertase Activity Units)
- lactase – LacU (Lactase unit) or ALU (Acid Lactase Units)
- lipase – LU or FIP (Lipase Units)
- maltase – DP (degrees diastatic power)
- malt diastase – DP
- pectinase – PGU or Endo-PGU (Endo Polygalacturnonase Units)
- protease – PC, HUT (Hemoglobin Unit Tyrosine base), or USP
- papain protease – FCC PU (Food Chemical Codex Papain Units)
- acid fast protease – SAPU (Spectrophotometric Acid Protease Units); this is the same as acid stable protease (SAP)

The higher the activity number, the quicker the food is digested. A lower number will still be digesting food, but it will take longer. Since enzymes do not get used up in the process, we do not 'run out' of enzymes before all the food is digested, BUT the stomach and intestines are absorbing food, completely broken down or not, at the same time. Since we are on the clock with possible unbroken-down peptides, sugars, or other food components being absorbed, we want the food to be digested by the enzymes *before* it gets absorbed in a partially broken-down state.

Example: Say Product A has 15,000 HUT of protease and Product B has 45,000 HUT of protease. Product B can break down three times more protein than Product A in a given period of time. This is how to compare digestive enzyme activity and formulations.

Part 4. Compare pricing – Calculating cost comparisons

Once you have picked a product that contains the enzymes you need to meet your goals, and you see that the label lists certified activity units, you have several ways to further compare products.

What is the cost per capsule?

To find out what the cost per capsule is, first find out how many capsules are in the bottle from the label. Capsules are better because the process of making tablets is hard on enzyme integrity or activity. Write this number down as Number of Capsules (per bottle). Next, add the price for the bottle, any extra discounts, taxes, and/or shipping charges to find the Total Cost (per bottle). Now divide the Total Cost by Number of Capsules. This gives you the Cost per Capsule.

What is the activity per capsule?

Sometimes it is helpful to compare products by activity per capsule. The product label may already list the activity per capsule. You always want to buy a product that lists the ingredients by acceptable units for activity, not by weight (such as in milligrams – mg). Weight tells you nothing. You can have 100 mg in an enzyme capsule but if it has zero activity, it is worthless to you. 100 mg may contain 5 units of activity or 500,000 units of activity. You can only compare values for activity if the units are *identical*. FCCLU is not the same as LU.

    If the units are not identical, you need to find the conversion factor to get them into similar units. Here are some of the known conversions. If the units are not identical and there is no conversion factor, you cannot make a side-by-side comparison and will need to look at other factors.

    1 FCC LU = 8.4 LU (units for lipase)
    1000 FIP = 10,000 LU (units for lipase)
    1 GAL = 2 AGSU (units for alpha-galactose)
    10 GDU (gelatin digesting units) = 15 MCU (milk clotting units)
    30 DU = 1 AG (units for amylase)
    100 BTU = 1,956,522 FCC PU (units for protease)
    100 HU = 88 HUT (units for protease)
    160 SU = more than 20 IAU (units for sucrase)
    chymotrypsin 1 mg = 1000 USP units
    trypsin 1 mg – 25,000 USP units
    papain 1 mg = 6000 USP units

The U.S. Pharmacopeia (USP) helped by establishing a standard for the pancreatic enzymes (animal-derived) by which you can compare other enzyme supplements, such as plant- and microbial-derived. This standard is called 'X' and each X contains an equivalent of:

25 USP units of amylase
2 USP units of lipase
25 USP units of proteolytic enzymes

If a supplement contains 5X pancreatic enzymes, it would provide five times the amount of each of the enzymes in this standard, or 125 USP amylase, 10 USP lipase, and 125 USP protease. There is no direct conversion between USP units and FCC units because they are produced from different sources, using different methods.

How many capsules will you need to take?

Compare how many capsules of Product A to Product B to get the same amount of enzyme activity. For example, let's say we have Product A, which lists one capsule as containing 12,000 units of amylase and Product B saying one capsule contains 18,000 units of amylase. We will need to take one and a half capsules of Product A to equal the activity of Product B. Now compare the costs per capsule to see whether one and a half capsules of Product A are more or less expensive than one capsule of Product B. If Products A and B cost the same amount, then Product B is the better buy. However, if Product B is twice as expensive, then Product A is the better buy.

You can also calculate the cost per unit of activity and then compare which is more expensive. Sometimes, the difference is not significant in terms of everyday use, but other times it can be very significant. In the example below, Product B gives you more amylase activity per dollar.

Example:
Product A has 12,000 DU amylase and costs $20 for 90 capsules.
12,000 DU units of amylase/$20 = 600 DU units of amylase/dollar

Product B has 18,000 DU amylase and costs $25 for 90 capsules.
18,000 DU units of amylase/$25 = 720 DU units of amylase/dollar

Product C has 20,000 DU amylase and costs $30 for 90 capsules.
20,000 DU units of amylase/$30 = 667 DU units of amylase/dollar

Another example:

Let's say I can choose from several enzyme products from different companies. They all contain the enzymes I want and they are all using approved activity units. I want to figure out the cost per capsule. One of the products comes as a powder instead of in capsules like the other products. I will say it is called Foodase and sold as a powder at $36/60 grams with one-fourth teaspoon = 1 gram. I need to determine how this product compares by starting with the same volume or amount of product as the quantity I am looking at in the capsules of the other products. The package of Foodase says to use one-fourth teaspoon per meal as a 'serving' with 60 servings per package. I open capsules of all the other enzyme products into a measuring spoon. The other capsules all equal about one-eighth teaspoon. Then I try packing as much Foodase into one of those capsules as possible, then empty the contents into the measuring spoon. I actually did this several times. The amount of powder equaled about one-eighth teaspoon. So, one-fourth teaspoon of Foodase equals two capsules. Or 1/8 teaspoon = 1 capsule. Or 1 package of Foodase = 120 capsules.

Consumers need to always compare the dose size with the ingredient per activity units. ALWAYS check how many capsules count as 'one serving' or 'one dose.' Just looking at the list of ingredients and the amount of activity on a label and automatically thinking it is for one capsule is very easy to do. Marketing departments know this too. It may look like you are getting a lot of enzyme activity per serving, however the serving size may be more than one capsule. This is a common practice for many dietary supplements, not just digestive enzymes. Also, note how many capsules per bottle. Usually for enzymes, capsules come in increments of 60, 90, or 120.

Now that I know that one-eighth teaspoon of powder = one capsule, I can compare the activities and costs for one capsule of each product. At this point, I will focus on calculating the cost per capsule and not activity. All costs given below are in cents per capsule. You may want to add in the total shipping costs, taxes, discounts, or whatever else to make sure you are dealing with the total cost. These fictional products and amounts are for the purposes of example only.

After you calculate the total cost of each capsule, then add all the capsules you will need for one meal together, and then add up the total number of capsules (cost) for the frequency you intend to use enzymes for. For example, a nickel more may not be much if you only plan on giving enzymes occasionally, but if you are planning on giving them for every meal and snack then it can add up...especially if you have more than one person in the family on enzymes regularly.

Company A
Prozyme Protein $22/90 capsules = 24.4 cents/capsule
Prozyme All-purpose $19/90 capsules = 21.1 cents/capsule

Company B
Digestase Alpha-protein $44/180 capsules = 24.4 cents/capsule
Digestase Beta-carbs $23/120 capsules = 19.1 cents/capsule

Company C
Foodase Proteins & Sugars $36/120 capsules = 30.0 cents/capsule
Foodase Fats $25/90 capsules = $27.8 cents/capsule
Foodase Starches $15/60 capsules = 25.0 cents/capsule

Now compare possible combinations and their corresponding costs: If you want a complete enzyme program for all foods

from Company A:
24.4 Prozyme Protein
21.1 Prozyme All-purpose (contains some protease)
___
45.5 cents total using 2 capsules for all food groups

from Company B:
24.4 Digestase Alpha-protein
19.1 Digestase Beta-carbs
___
43.5 cents total using 2 capsules, but have no enzymes for fats and sugars

from Company C:
30.0 Foodase Proteins and Sugars
27.8 Foodase Fats
25.0 Foodase Starches
___
82.8 cents total using 3 capsules for all food groups

You can mix and match enzyme products, so you do not have to buy everything from one company. Check to see that all the products you are considering use a very high standard of manufacturing methods and quality control. All products should use quality ingredients, including sulfite-free papain or a manufacturing method that does not require sulfite. This will not necessarily be listed on the label, so double-check from others or the manufacturer. You may want to note if products come packaged in gelatin capsules (animal based) or veggie capsules (vegetable based), if this is a concern for you.

You will also need to factor in individual responses. Any individual may have a better reaction to one formulation, but not another for some unidentifiable reason. Nothing in the lab, whether kitchen testing or real laboratory, or what other people with similar symptoms say, can always predict how any individual will respond.

Part 5. What other stuff is in the product besides enzymes?

You will also need to check for any additives, fillers, binders, or other ingredients that are in the product besides enzymes. You will have to decide if you need these extras at all, or want to pay for them, or if they may cause an adverse reaction. Possible items are:

- probiotics

- vitamins

- minerals that may help deliver or transport enzymes (calcium ascorbate, magnesium citrate, zinc, or manganese gluconate)

- amino acids

- other stuff: herbs (such as aloe vera powder, ginger root), whole foods, gelatin, additives, preservatives, colorings, dairy, soy, yeast, gluten, sugar, salt, corn, wheat, or hydrogenated oils

- potential allergens or food intolerances

- ionic minerals – these minerals *may* help the digestive enzymes become two to three times more active and effective

Part 6. Research the product and manufacturer

This is always a good idea. Call the company or manufacturer and get answers directly. Keep in mind that a company will usually want to paint their products in the best light possible. Usually on health issues, they will not deceive you or lie, although this is not guaranteed and does happen. Most probably, they may not be forthcoming in giving you all the information you need if that information may dissuade you from purchasing their products. Bluntly ask them to explain why you should buy their product over a competitor's product. This is not being pushy; it is being practical. People who are proud of their work are very happy to talk about it. If their products do not list FCC units, ask for the corresponding values. Have them explain it to your satisfaction. Be cautious about extra things in the formulation that you do not necessarily want to pay for.

Then go to one of the best sources available for information: Ask others if they have any experience with the products. Ask about side-effects and interactions. Find other individuals that have symptoms or a condition similar to your situation. Although parents and other adults have their preferences, they are usually very honest about that. Asking several individuals will give you a much better idea of general satisfaction with the product. In the end, you are paying for it and your family will be using it. Many issues surround the quality of enzymes. Ask about handling, storing, and packaging of enzymes because these all affect enzyme activity. We are interested in the activity of the enzyme as we ingest it, not as it leaves the factory.

Part 7. Understanding enzyme names and activity

Enzymes usually have the ending 'ase.' Usually, the first part of the name tells something about what the enzyme is working on. A protease would be an enzyme (ase) that works on a protein (prote). A lipase would be an enzyme (ase) that works on lipids (lipids means fat, so this enzyme breaks down fats). Pectase is an enzyme (ase) that works on pectins (a compound found in some fruits such as apples).

Protease is a broad term referring to any enzyme that breaks down proteins. In the enzyme business, almost all enzymes from microbial/fungal organisms are actually mixtures (or blends) of many different enzymes. For example, you can get a number of 'proteases' available from enzyme brokers with names such as 'protease 3.0', 'alkaline protease', 'acid-stable protease', or 'protease 4.5'. The enzyme blend called protease 3.0 may also contain bits of amylase, pectinase, and a variety of different peptidases. However, the supplier only certifies that blend for units of protease 3.0 which has certain characteristics that make it different from other proteases. These characteristics may refer to a number of things, including its optimum pH or particular affinity to specific substrates.

There are a few main companies in the United States that produce core basic enzymes. People making formulations buy what they want, similar to buying ingredients from the store and then cooking something special from them. Okay, let's say a manufacturer purchases three blends of a supplier's proteases, such as 'A', 'B', and 'C.' Then he mixes two parts of A with six parts of B. Now the manufacturer has a 'distinct and proprietary' blend, which he decides to call 'Ultrazymase,' and puts it on his label so that other sneaky manufacturers cannot copy his remarkable formula. The problem is, how do you then convey the activity of Ultrazymase? This explains why sometimes you do not get an exact ingredient list – because it is the proprietary information of the enzyme formulator. Or it may contain a proprietary blend from the original supplier, which even the formulator does not know exactly, or is not at liberty to disclose. It also explains why you may see a name that sounds like an enzyme because it ends in –ase, but you cannot find it in any research book or with a search engine on the Internet. These are usually the created names of proprietary blends.

It may seem logical to add as many different proteases in a product as possible to get the widest amount of proteins broken down; however, going with, say, three or four different proteases may do as much protein breakdown as having smaller amounts of six or seven different proteases.

Enzymes in Action
— Seven Months of Real-Life Use

Trends and responses and patterns and information were tracked on the Enzymes and Autism site for seven months, from May 13 to November 2001, summarized, and written up. Although many more people were using enzymes, this report details the results of 260 volunteering individuals. This section presents some of the information from the seven months that is not given elsewhere in this book. For the total in-depth report, please contact the author. The conception, development and writing of this summary report were completely independent of any manufacturer, and is intended as a guide to help parents and other individuals make better decisions with digestive enzymes with their families and others in their care. This represents enzyme therapy conducted by real people in their everyday lives.

By the time you read this book in print, it will be well over two years. That we have this information at all is due to the cooperative efforts of all the individuals supporting one another on this journey. The information presented is relevant to the products Peptizyde and HN-Zyme Prime (Zyme Prime) by Houston Nutraceuticals, *www.houstonni.com* which were available at the time. Keep in mind

Cheryl

that this information may or may not apply to other enzyme products due to the fact enzyme formulations can be quite different, as is each person's individual biochemistry and situation. The Enzymes and Autism group is an independent forum for families interested in the use of digestive enzymes to assist with nutritional issues related to autistic spectrum and related conditions. The group is located at:

http://groups.yahoo.com/group/enzymesandautism

or see the link at *www.enzymestuff.com*

[handwritten note in left margin: Internet group with arrow/star]

Executive summary

The bottom line is that the overwhelming majority of individuals using these enzymes saw significant improvements with Peptizyde and HN-Zyme Prime over a wide range of areas in a relatively short amount of time. Of 260 total individual cases (100 per cent) using these products for at least three weeks, 235 (90 per cent) reported positive results, 14 (6 per cent) reported negative results, and 11 (4 per cent) reported inconclusive results.

- Significant improvements were seen in eye contact, language, humor, foods tolerated, foods accepted, sleep, weight gain or loss, digestion, stools/bowels, overall appearance, transitioning, socialization, awareness, problem solving, short-term memory, flexibility in routine, range of interests, sound and light tolerance, sensory integration, spontaneous affection, and energy level, among others. Significant decreases were seen in aggression, hyperness, anxiety, self-stimming, self-injurious behavior, pain, and headaches, among others.
- Most positive results were apparent within the first three weeks.
- Most people say any side-effects or adjustment effects that start with enzyme use were manageable and ended within three weeks.
- The majority of respondents (about 85 per cent) were either 100 percent casein-free, gluten-free or partial casein-free, gluten-free with enzymes. More gluten and casein foods were added as people continued with Peptizyde over time. More foods were added in general with both products over time. People saw significant improvements whether they were on or not on a restrictive diet.
- After seven months on enzymes, there have been no reports of regression after success was seen past the three-week adjustment period due to enzymes.
- Improvements continued over time with continued enzyme use. The longer on enzymes, the more foods people tend to reintroduce

and the more positive results are seen. This could be due to better nutrition from a wider variety of foods being better digested, as well as progressively improved health over time.

- Most people return to eating most foods, although not all people can eat all foods, with these enzymes.
- It became apparent that besides matching the right enzymes to the right foods, timing and dissolution rate of the capsules was key to getting good positive results with enzymes. Some parents only saw consistent improvement when they adjusted the timing.
- Many people see the Happy Child Effect, which refers to the general, overall positive disposition of the child once they start taking Peptizyde and Zyme Prime regularly. The child becomes noticeably more pleasant, easy-going, cooperative, and helpful.
- People having difficulties or concerns when starting enzymes were far more likely to achieve positive results when they posted their concerns on the Enzymes and Autism message board for assistance. This is attributed to getting more education and information on using enzymes, which helped them learn about possible needed adjustments.
- Some people were able to significantly reduce their costs on food, supplements, medications and other therapies by using enzymes. This correlated to length of time on enzymes and to what extent they remained on a restrictive diet. Some medications could be reduced and others discontinued altogether.
- Most parents found using enzymes to be far more convenient and flexible than a restrictive diet. This greatly reduced stress, and increased happiness and quality of life for the entire family.
- A few people have seen no results either way; however, a few people have improved to the extent they no longer fit the criteria for their previous diagnosis. Most people said that Peptizyde and Zyme Prime were an important part of their child's or their own improvement but not to the extent that all other therapies or medications were no longer needed. However, at times, a restrictive diet could be eliminated and other supplements and medications either greatly reduced or eliminated. Most say that adding these enzymes was definitely beneficial and made most other therapies more productive. The people taking these enzymes say the addition of enzymes allowed them to streamline their overall plan to a more efficient, easier, convenient, cheaper, effective, and simpler treatment.

Summary of results of seven-months Peptizyde and Zyme Prime
One summary was compiled and written covering the first four months. Later, another one was created of the following months, the four-to-seven-month period. The information is listed separately where applicable and otherwise reflects the total seven-month period.

Terminology used in this summary
The information in this summary report is presented as general groupings because individuals used their own words and no standardized form was used. Keep in mind that not everyone commented on all subjects. The scale given on page 176 was used throughout this summary. This method of categorizing results is meant to convey the general idea of what is taking place and is not meant to be quantitative. Each person is an individual and responds differently. Also, the goals for each family are different, and so what they were hoping to achieve with these enzyme products varies.

I wondered in the beginning how 'acceptable' this type of rating scale would be. Finding an objective rating system to measure subjective results is always a challenge. For example, one mother may say, 'My son has improved dramatically with language.' But what exactly does 'dramatically improved' mean? Would that be the same level of improvement as another parent who states, 'My daughter is doing sooooo much better we can hardly believe it. She talks all the time now.' We can at least agree that both parents saw remarkable improvement, and it was very noticeable to them.

We have another situation of comparing very noticeable improvement in the same area, but the improvements in this area may be expressed quite differently. Using language again as an example, one parent says, 'My son has improved remarkably. He talks with words now instead of grunts and noises. He says things like, 'No want orange juice. Want grape juice.'' Clearly, this child has started not only talking but also communicating his preferences. Another parent may post a message saying, 'My daughter has improved remarkably. She no longer just blurts out her demands such as 'Juice. Give me grape juice.' She participates in long pleasant conversations with us. She asks others what they would like and then she gets it for them. She will talk about what she sees in a mall or what happened at school, or how she feels. She reads books to her little sister.' This child already had clear speech but she has progressed far beyond that into much more fluent and descriptive language. The progress in each case is great, even though the starting point was different. Age of the child is a factor too. These factors were taken into account when deciding on how much success was seen.

This is very common when trying to measure changes in behavior. Other areas of research accept as valid and employ very similar types of rating scales when evaluating the effectiveness of treatments involving behavior, moods, overall disposition, well-being or going by subjective observation. So how accurate are the parents' observations? How does this compare with a clinical study where professional researchers would observe the children? Parents' observations are usually much more accurate in judging these behaviors and rating them than professional researchers because they know their children best. They have been around their children far more and can recognize changes in patterns better than a researcher who may only watch a particular child for a few minutes, or maybe an hour or so.

A study was conducted aimed at measuring the reliability of parents' observations compared to researchers' observations. The results showed that the parents were significantly better at objectively and accurately rating improvements in their children (but as parents, we knew that all along!) (Herlihy 2001).

This type of summary report study is known as an epidemiological study and is valid in this regard. Epidemiological studies are different from the clinical trials or laboratory studies often conducted by companies, universities, and research organizations. Alot of people are interested in scientific studies in the hopes they will definitely 'prove' something one way or another. Epidemiological studies usually take place both before and after clinical trials. The ones done before clinical trials help determine if some significant situation exists that warrants a clinical study (like a pre-study). They are also helpful in determining how to conduct and measure a further scientific study. For example, in the case of the Seven-month Summary Report, we would now know that, in general, a great many people with autism are responding positively to enzymes for whatever reason so this therapy warrants investigation. We also know individuals with celiac disease should not be included because this is a special subgroup that reacts badly to proteases with gluten. And we would be aware of the timing of the veggie capsules dissolving so all enzymes would need to be given out of the capsules or the waiting time allowed. This one factor could influence the results but have nothing to do with the actual performance of the enzymes themselves.

The Seven-month Summary Report is also representative of what takes place in the last phase of testing, usually called 'field testing.' This is a very important step, and often the most important, because what you see in a test tube, lab, or clinic may not turn out to be practical in everyday life. Each type of study has its own role to play in the overall picture, and is valid and quite necessary. Much of the findings and guidelines produced from these typical families will no doubt be very helpful to those conducting further research. In return, scientific studies may shed light on why certain things appear in real, everyday life. Then, that information may help improve and further refine guidelines. The final goal for everyone concerned should be to contribute to giving parents, adults, researchers, and practitioners the best information possible so they can make the best decisions to improve the health and life for any particular individual.

Profile of individuals in the report

Individual goals – Parents were asked to select the symptoms they were hoping the enzymes would help with, or goals they hoped to achieve, from a list provided. This volunteer poll allowed individuals to select more than one choice. The results follow in order from most votes to least votes: absorb more nutrients from food and supplements; reduce stimming, tantrums, crabbiness, or other behavior symptoms; heal leaky gut; bowel problems – constipation, diarrhea, irregularity ; improve eating – self-limited foods, eats too slow or fast; be able to leave a restrictive diet; be able to have planned infringements on a restrictive diet; just catch accidental infringements while on a restrictive diet; reduce rashes, red ears, other physical symptoms; reduce nausea and reflux; nothing specific – just looking for general improvement

Diagnosis – Most of the people who responded characterized their children or themselves as High Functioning Autism, Asperger's, or PDD (some type of pervasive development disorder). Only a few described their children or themselves as moderate or low functioning. Some said their children or themselves had some other primary condition (including a variety of autoimmune conditions, asthma, allergies, sensory integration dysfunction, AD(H)D, or poor digestion).

Approximate age of individuals –
   (0–4 month summary, 151 people)
0–1 yr none
2–3 yr some
4–6 yr many
7–10 yr many
11–18 yr few
18 + few (parents caring for adult children)
adults some (parents or adults caring for themselves)

   (4–7 month summary, 109 people)
0–1 yr none
2–3 yr 18%
4–6 yr 29%
7–10 yr 25%
11–18 yr 5%
18 + none (parents caring for adult children)
adults 5% (parents or adults caring for themselves)

Diet Type – Trying to determine what type of diet the individual using enzymes was on became a little tricky because often this changed as they continued to use enzymes over time. Someone may have said they were 100 percent casein-free, gluten-free and also eliminated soy and corn, but after a couple months on enzymes, they may have reintroduced casein and corn. A few people wanted to try enzymes before trying a restrictive diet. Some started on a 100 percent restrictive diet but with increasing improvement with enzymes drifted to a less restrictive diet. The longer people were using enzymes, the more likely they were to reintroduce previously restricted food (of any type: casein, gluten, soy, corn, fruits, eggs, oils, etc.).

Others started both enzymes and a restrictive diet at the same time. (Starting more than one therapy or supplement at a time is generally discouraged because then you cannot accurately tell which therapy may be helping or hurting. If someone could not attribute the improvements or regression clearly to enzymes, this information was not included, except for any special circumstances given noted as such.)

In general, most people started with a strict casein-free, gluten-free, plus other food-free diet including removing artificial chemical or yeast-free. Most people continue to use enzymes plus some food elimination together for a solution that is workable and best for their family. The longer people were using enzymes, the more likely they were to reintroduce previously restricted food (of any type: casein, gluten, soy, corn, fruits, eggs, oils, etc.).

In the second collection of information, more people said they were not on a restrictive diet 100 percent. Besides the above considerations, another possibility that more reported using enzymes 'off-diet' at least sometimes was because the discussion of intentionally using enzymes with casein and gluten food became more accepted. Prior to the introduction of these enzymes, there was much alarm that anyone would intentionally consume casein or gluten because of the rigorous demand that you needed to be 100 percent casein-free and gluten-free at all times. Some individuals did not feel free to speak freely of their actual situation because of the admonishment they might receive.

Enzymes used

Using both enzyme products provides enzymes to break down all food types. Some used Peptizyde only. Most parents who started with just Peptizyde eventually added Zyme Prime. This was usually due to learning more about all the other benefits of enzymes, and finding that using enzymes for all food groups gave better overall results. It was common for parents to adjust the doses and types of enzymes as they determined which foods to use with which enzymes. The reasons cited for using Peptizyde include:

- to break down the casein/gluten/soy proteins for overall improved digestion (most)
- to have planned infractions (many)
- to catch any unknown or contamination sources (many)
- to reintroduce casein/gluten foods in sensitive individuals (some)
- to improve energy levels (few)
- to improve immune system (few)

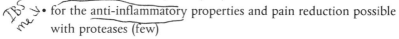

- for the anti-inflammatory properties and pain reduction possible with proteases (few)

The reasons cited for using Zyme Prime (broad-spectrum enzyme product) include:

- to break down other foods besides casein/gluten/soy foods for overall improved digestion (most)
- to reintroduce other foods besides casein/gluten that an individual may be sensitive to – such as, fruits, corn, potatoes, tomatoes, nuts, eggs, chocolate, etc. (most)
- to achieve the 'Happy Child Effect' (many)
- offset hyperactive behavior (some)
- to improve energy levels, immune system, or other concerns (few)
- to improve immune system (few)

Results with Peptizyde and Zyme Prime

Each person trying enzymes in a family counts as one 'individual'. If a person saw negative results, but two weeks later said it became positive, this counted as a positive result, and vice versa. In each case, the person who posted determined what they would call their final result for each person in the family. If one member of the family could not tolerate the enzymes but two others could, this counted as one negative and two positives. Results were parent-judged and included results based on physical symptoms (such as red ears or bottom, stool condition) as well as behavioral reactions (such as tantrums, screaming, lethargy, social withdrawal, eye contact).

Percentages (0–4 months)
100%   151 Total number of individuals using enzymes
 87%   131 Number of those with positive results
  8%   12 Number of those with negative results/no results at all
  5%   8 Number of those with inconclusive results

Percentages (4–7 months)
100%   109 Total number of new individuals using enzymes
 92%   100 Number of those with positive results
  5%   6 Number of those with negative results/no results at all
  3%   3 Number of those with inconclusive results

Total Percentages (0–7 months)
100%   260 Total number of new individuals from both summaries
 90%   235 Number of those with positive results
  6%   14 Number of those with negative results/no results at all
  4%   11 Number of those with inconclusive results

When calculating the total of the seven months, several people who initially reported negative results followed the guidelines for getting best results, and then saw positive results and so switched categories.

Of the inconclusive results:
All individuals in this category stated themselves they had inconclusive results and wanted to try again later. Most families saying they had inconclusive results stopped using the enzymes for some other reason not related to a reaction to the enzymes themselves, including:

- The person had mixed positives and negatives and wanted to continue to work on it
- The child went to a relative's house where the relative was not supportive of the diet and/or supplements
- The child started a new therapy and family wanted to concentrate on that only
- Several supplements or therapies were started at the same time and so the parent could not determine which supplement or therapy was a problem or helping

Of the negative results:

Negative posts were those where individuals said the results were negative or included such descriptions as: failed, disappointed, not working, discontinued etc. The following reasons surfaced:

- Two people saw no change at all either positive or negative, and counted themselves as negative
- One person had an allergic reaction (upper body hives to papain and/or bromelain)
- Two people had continued excessive negative hyperactivity
- Six parents reported their child became unacceptably aggressive and hyperactive when on Peptizyde alone. Behaviors include hitting, pushing, pinching, throwing objects, arguing, and throwing self onto the floor or furniture. Although we have a few recommendations for dealing with this, these parents opted to discontinue enzyme use and are counted as negative results. These types of reports decreased almost completely once the L-glutamine was removed from the formulation in September 2001.
- Five parents said their child 'could not tolerate' the enzymes and 'reacted' negatively with nothing more specific. Because the enzymes were used less than three to four weeks in these cases, it is not known if this includes a side-effect that would resolve itself over time or not.
- Three parents said they discontinued the enzymes due to persistent diarrhea or constipation with enzymes

Of the positive results:

The positive results are described in Chapter 12, The Happy Child Effect. Positive results were interpreted to mean that the individual showed noticeable improvement and/or that the parent achieved whatever individual goals they had with enzymes.

- Many (around 35 per cent) described the level of improvement as very great or immense, using such terms as: immense improvement, amazing, incredible, overwhelming, unbelievable, awesome, etc.
- Many (around 35 per cent) described the level of improvement as very good or good, using such terms as: noticeable,

significant, very good, much more, great, surprising, unexpected, etc.

- Some (around 20 per cent) described the level of improvement as good to okay, using such terms as: steady improvement, slow but constant, somewhat, progressively, some, etc.
- Few (around 5 per cent) described the level of improvement as positive overall without any huge immediate gains, using such terms as: somewhat, incremental, little bit, sort of, I think there is, and also included those who saw improvements in physical symptoms but not in behavior, or in behavior but not in physical symptoms (or at least to the level and range they were hoping for). This indicates that most people saw positive results, and when they did see positive results it was in a significant and noticeable way

Side-effects and adjusting to enzymes
Please see Chapter 9, What to Expect When Starting Enzymes.

Effect of age
This is discussed in Chapter 19, Enzymes and Restrictive Diets.

Possible Reasons for Improvement in Results. The second group of individuals (109 people) saw a better percentage of successful results than the people responding from the first four months (151 individuals). The following factors contributed to the increase in positive results.

L-glutamine – The first formulation of Peptizyde contained L-glutamine which was suspected as the cause of some of this because it has excitatory properties. It was removed from the formulation. People using both versions reported a significant decrease in their child's hyperness if they saw any difference at all.

Having suggestions for excessive hyperactivity – As noted in Chapter 9, several reasons explain why hyperness may occur with the start of enzyme therapy. Depending on the individual, hyperness may be a positive improvement or a negative reaction. When considered a negative reaction, the following suggestions proved most helpful.

- giving magnesium to correct a deficiency, help with calming
- giving Epsom Salts to provide sulfur for processing phenolic compounds and magnesium
- adding Zyme Prime
- decreasing or discontinuing other supplements

Better information and allowing three weeks for adjustments – After the first four months, it became apparent that you need to allow a three-week period for the body to adjust to enzymes, or withdrawl effects. Now people knew to allow for that.

Although the Enzymes and Autism group was started as a place of education, information, and support for parents interested in using enzymes, there was an interest in whether this factored into the success of enzyme use or not. It seems that those posting or lurking may have had better overall positive results than those who did not participate. There was also a concern that the success rate for Peptizyde and Zyme Prime might be inflated over reality because:

- People posting were a self-selecting group
- People seeing negative results may just not want to post their adverse observations (even though this is always encouraged)
- Far more people are members of the enzyme group than post (although this is very standard for message boards). However, most people who subscribe do remain for extended periods of time even if they do not post. This indicates they are getting some information of value from being a member.
- Not everyone who uses Peptizyde and Zyme Prime belongs to Enzymes and Autism so the results posted may have differed than those of individuals who did not post

However, in general, it is a fact of human nature that most people are far more likely to voice or post complaints, problems, and gripes than speak out positively. Looking at the content of the posts, far more involved troubleshooting and assistance to resolve a problem or concern than actually have content of a positive nature. Considering this, the fact there are so many positive results posted tends to verify that you are most likely to see positive results. In an effort to gather additional information, and to test if people who do not post were actually seeing more negative results, 38 individual emails were sent out to members who had either posted saying they were interested in enzymes or posted they had started Peptizyde and Zyme Prime but had not posted any follow-up on their progress or observations.

Twenty-two messages were returned, about 58 percent. This was very good because most polls and marketing research only bring in about a 3 to 5 percent return. This high rate of return is probably because people are more likely to respond to a direct personal email from another parent on something in which they were personally interested. These numbers reflect the response from emails sent over the entire seven months because replies came in over that time. Of these 22 responses: 4 were negative (18%); 4 were inconclusive (18%); and 14 were positive (64%). Compared with the summary report results, this indicates that people not posting results or seeking suggestions tended to see less positive results than those who did. Non-positive comments included such things as:

- The parent had not really kept up with the enzymes
- Decided not to try enzymes
- Saw some problems such as hyperness and loose stools and interpreted this as a negative reaction so stopped

The positive replies were very similar to those posted and included:

- The person does not usually post much (not their style)
- Was meaning to but didn't get around to it
- Everything was going well and only post when they have a problem
- The child was doing well, they have all the information and so do not frequent the site anymore

To see if the suggestions usually given from the group would be helpful, I sent replies to those people who said they had some problems and had asked for suggestions (if someone said they were not successful and/or did not ask for help, none was offered). Eleven emails were sent offering the following information which is frequently given on the message board, seven to those who were seeing success already but asked for further assistance and five to those who indicated negative or inconclusive results so far:

No fatigue?

- Loose stools in the beginning – common with any enzyme; should end in a few days
- Stomach aches – Stop proteases for four to five days and then resume. Most people have no trouble after that.
- Hyperness – Very common; try the non-L-glutamine Peptizyde, give Epsom salts, and/or add magnesium to the diet. Also, adding a broad-spectrum enzyme helped decrease hyperactive.

Four replies were received from those already seeing positive results and four replies were received from those who previously saw negative or inconclusive results all saying the suggestions were found to be very helpful and the person now saw much more positive results. This indicates that those seeking assistance from other people through a forum such as the enzyme group are more likely to see better results and successfully troubleshoot concerns and problems.

In addition, over 65 'threads' or conversations were tracked on the message board where a person posted a problem or concern and was given suggestions for resolving the problem. These proved to be helpful in improving the situation. Very few concerns remained unanswered or unresolved. The evidence shows that the Enzymes and Autism group significantly helped to boost the number of people seeing positive results with enzymes.

People having difficulties or concerns when starting enzymes were far more likely to achieve positive results when they asked for assistance. This was attributed to the education, information, and support they received from the group or others, which helped them make the needed adjustments for success. Enzymes are relatively new to many people, and once they have a better understanding of how they work in the body, the success rate among those having problems or concerns greatly increased. The time and scope of adverse side-effects and reactions decreased noticeably because people were more informed and had a better idea of how to introduce enzymes, what to look for, and how to interpret reactions. People could get 'up and running' more quickly and effectively.

Celiac disease – If celiac is a possibility do not give gluten.

Timing issue – By the second summary, it was well established that two factors are key in using enzymes: matching the right enzymes to the food, and ensuring the enzymes were in contact with the food.

At the writing of the seven-month report in late December 2001, there were 1065 members at the Enzymes and Autism group six months after being established. Of course, not every member commented on the same topics, behaviors, or improvements, or even at all. Periodically, members were repeatedly encouraged to send in any results – positive, negative, and no change at all. Knowing when something wasn't working and why it wasn't working was as important as finding out when it was working. When people left the group they received an email asking if they had a particular reason for leaving. Most everyone replied to this message with a reason of some sort. The overwhelming majority of respondents said there were too many messages (the site was too active for them) and it was not due to any specific enzyme results. Members were not leaving because of problems with the enzymes. Actually, there are more members leaving because the enzymes were working quite well, and they no longer needed the support or information. This is a working parental research group as well as providing support and education on enzymes. When problems arise or we see certain patterns emerging, the group tends to discuss, research, and contribute to finding a solution.

Another method of gaining information was by emailing members privately who said they started enzymes at least once or were considering enzymes. I asked them to comment on any type of results, even if they had decided against trying enzymes at all. My feeling was that all information was useful, including reasons why enzymes may not show improvements or why someone chose not to go in that direction. If there were reasons why enzyme therapy was not practical or conflicted with another therapy, that was also helpful to know in order to provide better guidance so each of us could make better decisions.

There is also the enzymestuff.com web site for information.

Frequently Asked Questions

1. What are enzymes?

Enzymes are proteins made by cells in our bodies and all living organisms. They are specialized proteins that do work, such as synthesizing chemicals, rearranging molecules, adding elements to compounds, and breaking down compounds. Enzymes are protein catalysts. They cause biological reactions to occur under conditions that sustain life. Many types of enzymes exist and each type does a specific function. For an enzyme to work, the material it works upon must be present. If no such material is available to the enzyme, the enzyme performs no function. There are also metabolic enzymes in our body, but this discussion is limited to the digestive enzymes that help break down food.

2. How do enzymes work?

In general, enzymes work as catalysts of biochemical reactions. A catalyst increases or accelerates the rate of a reaction. The thousands of chemical reactions that occur in our body every second could not happen without enzymes to speed up these reactions. For example, a protein can be broken down into amino acids in the lab without the use of an enzyme, but to do so requires extreme temperatures, high pressure, or very strong acids; conditions not compatible with life.

Even with these conditions, it often requires hours to complete the reaction in the lab. Enzymes, in this case a mixture of proteases, can complete this reaction within minutes in water at normal temperatures. Another unique aspect of enzymes is that they facilitate the reaction without being destroyed or changed in the process. Because of this, one enzyme molecule could theoretically change an infinite amount of substrate if given an infinite amount of time. Increasing the amount of enzyme decreases the time required for completing the process. If you double the number of enzyme molecules, the time for the reaction decreases by half.

3. Where do enzymes come from and where do they go?

Enzymes exist in all raw food. All raw foods, including meats, have some enzyme activity. For example, green bananas have amylase, an enzyme that breaks down starch to glucose. In a number of days, the amylase converts the raw starch of the banana to sugar, which is why darkened bananas are so much sweeter tasting. Kiwis have an abundance of a protease known as actinidin, which is why you cannot make jello with fresh kiwis, or fresh pineapple. The protease degrades gelatin protein such that it cannot 'harden' or set.

Cooking or other types of processing destroys enzyme activity. This is the basis for 'canning' of vegetables – the heat destroys the enzymes and this preserves the food. Food enzymes can survive the pH of the stomach (about 4.5 to 5.5) for some time and so can contribute to the digestion of food while in the stomach. Animals, including humans, produce the enzymes they need from amino acids.

The more raw food you eat, the less digestive enzymes your body needs to produce. You can also take enzyme supplements, which come from animals, plants or microorganisms. Your body may recycle digestive enzymes from any source until they wear out. Enzymes in circulation perform many other tasks that assist in restoring and maintaining good health. Eventually, when these enzymes wear out, other enzymes break them down and the body uses the component amino acids for other purposes. They may also be excreted.

4. Why would you want an enzyme product as a nutritional or dietary supplement?

Proper and complete digestion is essential for good health. Digestive enzymes, used properly, can provide a substantial benefit to most everyone, especially those consuming a great deal of cooked or processed food. Enzymes may ease bloating, gas and heaviness with meals. Proteases may be beneficial between meals if you have an autoimmune condition or are recovering from an injury or illness. Enzyme-based products are one of the least utilized dietary aids, despite the fact that their use dates back thousands of years. Many manufacturers may avoid the use of enzyme products because their mode of action can be complex, and requires specific scientific knowledge and expertise.

5. Why take an enzyme supplement if I can just eat raw foods?

You could, except that raw food diets are difficult to maintain. Eating raw meat, with the danger of bacterial toxins, is not advisable as a means of obtaining food enzymes. Also, the amount of enzymes present in raw food is such that it would take many hours to adequately digest the food, and raw food does not necessarily contain all the enzymes needed. Enzyme supplements are a concentrated form of food enzymes that you may add to any diet so breakdown of food will occur at a faster rate.

6. Aren't enzymes taken orally destroyed by stomach acid or the body's own enzymes?

Some enzymes taken orally, including those made from animal pancreas extracts, become inactive in the low pH of stomach acid. Microbial-derived enzymes are acid-resistant, and can maintain activity at pH as low as 2.0 and as high as 10.0. Microorganisms use their enzymes to break down and digest the plant material that they grow upon. Since the site of fungal growth in nature can vary, fungi have evolved enzyme systems that allow the plant to grow under a variety of conditions, including differences in pH and temperature. Pancreatic enzymes work under a much narrower range of pH in the animal, since their environment is more controlled.

Manufacturers often enterically coat pancreatic enzymes to provide resistance to acidity. Some of these coatings contain ingredients considered unhealthful.

7. Can proteases be dangerous?

Used appropriately, these enzymes pose no danger to the consumer. The biggest problem with enzymes is the inhalation of large quantities of enzyme dust by people who work around enzymes a great deal, or getting large quantities on one's skin, such as in manufacturing plants. In both circumstances, enzymes can cause irritation, itching, and discomfort. It is rarely a problem with normal digestive use.

8. Will enzymes digest my mouth, stomach, or intestines?

If this were true, the enzymes produced by your own body would have already digested you away. These enzymes much prefer the denatured (cooked or damaged) proteins found in foods. Most proteins, in their healthy state, are coiled and globular in structure. This prevents the enzymes from having access to cleavage sites. When heated or in extremes of pH (like stomach acid), the proteins uncoil, exposing sites where the enzyme can bind and cleave.

Also, the cells of our bodies and the mucosal lining of the gastrointestinal tract contain protease inhibitors that inactivate certain protease enzymes. The mucosal layer acts as a physical barrier to proteolytic activity on living cells. The pancreas exposes your small intestine to a barrage of enzymes every time you eat, and the enzymes do not break down the intestine. If enzyme powder remains on the lips or gums for a prolonged time, it may break down some of the layer of dead cells that covers our mouth and throat (the whitish-colored layer). When the fresh, raw layer of tissue comes into contact with saliva, the person may feel slight irritation. If you open a capsule to mix it with food, make sure you drink something afterwards to wash any enzyme residue down. Because amylase is in saliva, you have a constant supply of enzymes in your mouth all the time anyway.

9. Can one become dependent upon oral enzymes? Will the pancreas stop functioning if I take enzyme for a long time?

No. Research has shown some adaptability of the pancreas in animals; giving oral enzymes resulted in a slight decrease in pancreatic enzyme output that quickly returned to normal once supplementing with enzymes stopped. The pancreas does not stop functioning.

10. I have never used an enzyme product with my child. What kinds of reactions can I expect?

Reactions from enzymes can vary. If your child is not on a casein-free, gluten-free diet, the use of enzymes that break down casein, gluten, or any other 'addictive' substance may cause a withdrawal effect. The child may actually appear to be worse for a few days. This is because of the decrease in peptides in his or her system.

Symptoms may appear as hyperactivity, being more sensitive to pain and stimulus, aggression, and sleep disturbances. In addition, you may notice digestive changes, such as increased frequency of bowel movements (not diarrhea), less stool passed, and a possible increase in gas production. All these are temporary and should be resolved in a matter of days. An increase in urination and thirst may appear too.

Reactions
Fatigue

11. How long do these symptoms last?

Adjustment symptoms may not appear at all, but if they do, most people get through this period in about a week. Most people find the symptoms are very mild, manageable and end by the third week. In a few instances, the symptoms may last longer. Please see Chapter 9, What to Expect When Starting Enzymes.

12. What about dosing? How much do I need to give?

Enzyme dosing is by the amount and type of food eaten and not based on age or weight. A good plan is to start with part of one capsule of one product per significant meal or snack. Then increase to one capsule, then start in a similar way with other enzyme products. In the neuro-typical population, there is no established upper limit on enzyme dosing, and no toxicity associated with oral enzymes.

How many capsules or tablets would depend on the specific formulation and the individual's need. For those individuals who tend to 'graze,' interval dosing may be more appropriate. Most people need to experiment (within reason!) with dosing to find the level that best suits you or your child's individual needs. See Chapter 11 Guidelines for Giving Enzymes.

13. When is the best time for giving the enzymes?

Preferably with the first few bites of the meal or just before mealtime. If you forget, go ahead and take the enzymes during the meal and even at the end of the meal. The important point is to take the enzyme, which works on contact. Food stays in the stomach for up to 90 minutes, therefore, introducing enzymes anytime during the meal will still provide benefits. Some people see much better results if they wait 20 to 30 minutes after swallowing enzymes contained in vegetable-based capsules before they eat. Apparently the vegetable capsules take longer to dissolve in the stomach environment. Waiting before eating allows time for the capsule to dissolve and sufficiently release the enzymes into the stomach.

14. My child will not swallow capsules. Can I sprinkle the enzymes on her food?

Yes. The enzymes will mix with food or beverages. Just pull the capsule open, and empty it out. If you use less than an entire capsule, you can just click the capsule back together and use the rest of the enzymes later. The taste and smell of enzyme products vary a great deal, which you may want to mask with fruit juice, ketchup, etc. Other products have no taste or smell at all. Please be careful to not inhale the enzyme powder, and to clear any residual enzyme powder from the mouth and throat area with additional food or beverage. Just make sure your child wipes their mouth after taking the enzymes. See the section on mixing suggestions in Chapter 11.

15. My child took an enzyme, then did not eat. Is this harmful?

No. If the substance is not present, the enzymes will not do anything. Since enzymes are protein, the body will eventually process the enzymes as any other food protein. Many people take high doses of enzyme products on an empty stomach to facilitate systemic enzyme therapy, that is, the uptake of enzymes into the circulation. This is common in treating various conditions with goals other than just the breakdown of food.

16. How can I give the enzymes with my child's school lunch?

You will need to check with your school on their policies for over-the-counter substances. See Chapter 20, Enzymes at School. You may try mixing the enzymes into a cold drink in a thermos and adding ice cubes to keep it cold. The cooler (even frozen) you keep an enzyme, the longer it lasts. This is important because enzymes become activated, and then start losing their activity once put into a liquid solution. Making chocolate wafers is another popular idea. See the section on mixing suggestions in Chapter 11 for recipes and other ideas.

17. How do enzyme supplements help in autism, other pervasive developmental disorders (PDDs), and neurology?

Location, location, location! A fundamental issue is the fact that there is an extensive nerve network (the enteric nervous system) running along the entire gastrointestinal tract. So anything that affects the gut directly affects the nerves. This leads to digestive enzymes having a direct impact on neurology. In addition, the largest part of the immune system in our body is in the mucosal lining on the frontlines of the gut. When working appropriately, the immune system makes sure any undesirable elements do not cross over from the gut into the body while allowing the good stuff through.

Many people having neurological difficulties also have general digestion problems, leaky gut, inflammation in the gut, yeast overgrowth, or other conditions that result in insufficiently digested food and poor absorption which enzymes may help improve. Some, not all, children with autism exhibit behavioral problems that lessen with the removal of certain foods. Enzymes help break down foods more sufficiently so they will not be in a form that is problematic or causes an immune system reaction. At the same time, they may help heal the fundamental gastrointestinal issues.

18. Peptizyde is considered the breakthrough. What makes Peptizyde so special?

Peptizyde contains a unique blend of proteases and peptidases targeting the identified problematic peptides, which made it the first product to allow some sensitive individuals to eat casein and gluten regularly. A recent research study confirms that the synergistic effect of these particular proteases are particularly effective in breaking casein and other protein bonds.

19. Can I use the enzymes in place of a restrictive diet?

Most people find they can reintroduce most foods by giving the appropriate enzymes for the food type. However, not all people can successfully reintroduce all foods. You may still need to restrict a few items on a case-by-case basis. Particularly problematic foods are nuts, seeds, and foods conveying a 'true' allergy. Enzymes can supplement any type of diet, and people show improvement in most circumstances. Remember enzymes are specific to the foods they break down, so you need to have the right enzymes for the corresponding food type. Many people find when they use certain enzymes they no longer need to follow a restrictive diet.

Enzymes are not a one-to-one equivalent for food elimination because although there is great overlap, they work in different ways. Enzymes very often accomplish much more than a restrictive diet. It also depends on the goals and purpose of the eating plan. For example, a diabetic follows a 'diet' but not for weight control. Or the Feingold and Failsafe programs include eliminating non-nutritious additives.

20. How do enzymes compare with a casein-, gluten-free diet?

Some have wondered whether enzyme products can replace a casein-free, gluten-free diet. If this is your goal, you need to look for a product specific and effective enough for this purpose. Many parents have been using Peptizyde successfully as an alternative to a casein-free, gluten-free diet for over two years and reporting that their children are showing immense improvement beyond what they saw on a restrictive diet, or even on a diet plus enzymes. Each person will need to decide for him or herself whether this is the best course of action for their situation. Supplementing this restrictive diet with enzymes not only helps to reduce or inhibit the production of any potentially harmful peptides, but also to support digestion and ensure complete degradation of food such that they may not be problematic, and increase the availability of nutrients at the same time.

21. If my child is already on a diet, do I still need enzymes?

There are a number of restrictive diets found to be helpful with various subgroups of autism spectrum, neurological conditions, and digestive problems – Feingold, yeast, Specific Carbohydrate Diet, casein-free/gluten-free, high protein/low carbohydrate, and others. Since enzymes facilitate food breakdown, absorption, and utilization they may be very helpful with any of these. The exact mechanisms of many problematic foods are not completely understood. Sources of peptide production from within the body, such as normal breakdown of red blood cells, yeast, and bacteria (good and bad) may be contributing to any peptide load. Or carbohydrate and sugar sources may be unknown. Or there is gut injury not accounted for. This may explain why some people do not see much improvement with certain restrictive diets, but do with enzymes. Adding enzymes to facilitate the digestion of what you do eat may make a diet more effective, such as adding proteases to a high protein diet.

Enzymes work very well to supplement a diet and thus make the diet more effective than it would be otherwise. Many diets are based on certain foods not being well-digested to begin with so enzymes are a natural solution. Enzymes work very well on actual food and natural food chemicals. The breakdown of artificially produced chemicals or added chemicals is somewhat limited at this time. These non-food ingredients usually do not provide nutrition anyway, may aggravate a sensitive system, and may hamper health in the long run. Enzymes are not able to convert non-nutritious compounds into healthful nutrients. Providing nutritious raw materials is necesary. Enzymes may enhance the absorption and ultization of any supplements or medications taken as well, making them more effective.

22. Do I have to get my doctor's approval?

Although enzymes are considered a safe food not needing a doctor's approval, it is always advisable to get the input of your health care physician when making decisions or changes regarding your health. This includes with any diets (particularly very restrictive ones), supplements, or over the counter products. Many supplements or diets can affect neurology. Some supplements may not be safe in higher amounts. A medical professional needs to know everything you are taking or would like to take in order to make good recommendations for your health. Although enzymes are very rarely a problem, adverse interactions are possible among other 'all natural herbs,' vitamins, or over-the-counter compounds. Most doctors do not have much training in nutrition and they may recommend you see someone who specializes in nutrition. Most doctors consider digestive enzymes a safe thing to try, just like you would bottled water, fresh fruit, or whole grains not really requiring medical approval.

23. What Else Do I Need to Know?
With the enzymes monitored, many people are able to significantly reduce their costs for food, supplements and therapies. Many families are able to find more manageable and enjoyable ways to assist their children and improve their quality of life. All age groups show improvements of some kind, even significant ones, including teens and adults, although results for any specific individual may vary considerably.

Enzymes have been used for years to safely assist in food intolerances and allergies, leaky gut, yeast, and immune system support among others. Enzymes help calm the neurological and immune systems down. Softly and steadily . . .reducing the total load. Reducing the hypersensitivity.

These well-researched benefits probably explain the overwhelming success rate and the wide range of improvement. Because enzymes have proven to help each of the wide variety of biological conditions seen with many neurological conditions, including autism, the high success rate of people improvement is logical, reasonable, and based in sound scientific principles. Of the individuals who tried these enzymes for other conditions, the vast majority have reported at least some benefits (digestive disorders, chronic fatigue, AD(H)D, asthma, fibromyalgia, sensory issues, migraines, allergies, etc.)

References

Alam, Z., Coombes, N., Waring, R.H., Williams, A.C., and Steventon, G.B. (1997) 'Platelet sulphotransferase activity, plasma sulphate levels and sulphation capacity in patients with migraine and tension headache.' *Cephalalgia* 17(7):761–764.

Alberti, A., Pirrone, P., Elia, M., Waring, R.H., and Romano, C. (1999) 'Sulphation deficit in 'low-functioning' autistic children: A pilot study.' *Biological Psychiatry* 46(3):420–424.

Ames, B.N., Shigenaga, M.K., and Hagen, T.M. (1993) 'Oxidants, antioxidants, and the degenerative diseases of aging.' *Proceedings of the National Academy of Science 90*:7915–7922.

Anderson, I.H., Levine, A.S., and Levitt, M.D. (1981) 'Incomplete absorption of the carbohydrate in all purpose wheat flour.' *New England Journal of Medicine* 304:891-892.

Autism Research Institute (2000) 'Parent Ratings of Behavioral Effects of Drugs, Nutrients, and Diets.' San Diego, California. ARI Publications.

Balakrishnan, V., Hareendran, A., and Sukumaran Nair, C. (1981) 'Double-blind cross-over trial of an enzyme preparation in pancreatic steatorrhea.' *Journal of the Association of Physcians of India 29*:207–209.

Bamforth, K.J., Jones, A.L., Roberts, R.C., and Coughtrie, M.W. (1993) 'Common food additives are potent inhibitors of human liver 17 alpha-ethinyloestradiol and dopamine sulphotransferases.' *Biochemical Pharmacology* November 17;46(10):1713-20.

Bartsch, W. (1974) 'Proteolytic enzymes in the treatment of herpes zoster.' *Der Informierte Arz 2*, 10, 1–7.

Beard, J. (1911) The Enzyme Treatment of Cancer and its Scientific Basis. London: Chatto & Windus (out of print).

Beazell, J.M. (1941) 'A Reexamination of the role of the stomach in the digestion of carbohydrate and protein.' *American Journal of Physiology 132*:42–50.

Bengmark S. (1998) 'Ecoimmunonutrition: A challenge for the third millennium.' *Nutrition 14*(7–8):563–572.

Beshgetoor, D. and Hambidge, M. (1998) 'Clinical conditions altering copper metabolism in humans.' *American Journal of Clinical Nutrition* May 67(5 Suppl):1017S-1021S

Bhagavan, H.N., Coleman, M., and Coursin, D.B. (1975) 'The effect of pyridoxine hydrochloride on blood serotonin and pyridoxal phosphate contents in hyperactive children.' *Pediatrics 55*(3):437–441.

Billigmann, P. (1995) 'Enzyme therapy – An alternative in treatment of herpes zoster. A controlled study of 192 patients.' *Fortschritte der Medizin 113*(4):43–48.

Blonstein, J.L. (1967) 'Oral enzyme tablets in the treatment of boxing injuries.' *The Practitioner* 198:547.

Bonney, R.J., and Davies, P. (1984) 'Possible autoregulatory functions of the secretory products of mononuclear phagocytes.' In Adams, D.O., and Hanna, G.M., Editors. *Macrophage Activation* 13:198-219. New York: Plenum Press.

Boyne, P.S., and Medhurt, H. (1967) 'Oral anti-inflammatory enzyme therapy in injuries in professional footballers.' *The Practitioner* 198:543.

Brenner, A. (1979) 'Trace mineral levels in hyperactive children responding to the Feingold diet.' *Journal of Pediatrics* 94(6):944–945.

Bucci, L.R. (1995) *Nutrition Applied to Injury Rehabilitation and Sports Medicine.* CRC Press. Boca Raton, Florida.

Buie, T., Winter, H., and Kushak, R. (2002) 'Preliminary findings in gastrointestinal investigation of autistic patients.' Harvard University and Massachusettes General Hospital. www.mgh.harvard.edu/children/dept/medical/ladders_doc.html.

Byun, T., Kofod, L., and Blinkovsky, A. (2001) 'Synergistic action of an X-prolyl dipeptidyl aminopeptidase and a non-specific aminopeptidase in protein hydrolysis.' *Journal of Agricultural and Food Chemistry* 49(4):2061–2063.

Cade, R., Privette, M., Fregly, M., Rowland, N., Sun, Z., Zele, V., Wagemaker, H., and Edlestein, C. (2000) 'Autism and schizophrenia: Intestinal disorders.' *Nutritional Neuroscience* March.

Carey, T., Ratliff-Schaub, K., Funk, J., Weinle, C., Myers, M., and Jenks, J. (2002) 'Double-blind placebo-controlled trial of secretin: effects on aberrant behavior in children with autism.' *Journal of Autism and Developmental Disorders* June 32(3):161-7.

Carter, C.M., Egger, J., and Soothill, J.F. (1985) 'A dietary management of severe childhood migraine.' *Human Nutrition Applied Nutrition* 39(4):294–303.

Chandra, S., Chandra, R.K. (1986) 'Nutrition, immune response, and outcome.' *Progress in Food and Nutritional Science* 10(1-2):1-65

Danicke, S., Simon, O., Jeroch, H., and Bedford, M. (1997) 'Interactions between dietary fat type and xylanase supplementation when rye-based diets are fed to broiler chickens 2. Performance, nutrient digestibility and the fat-soluble vitamin status of livers.' *British Poultry Science* December 38(5):546-56.

Dasgupta, M.K., Kenneth, G.W., Kaiviayil, V., and Dossetor, J.B. (1982) 'Circulating immune complexes in multiple sclerosis: Relation with disease activity.' *Neurology* 32:1000–1004.

D'Eufemia, P., Celli, M., Finocchiaro, R., Pacifico, L., Viozzi, L., Zaccagnini, M., Cardi, E., and Giardini, O. (1996) 'Abnormal intestinal permeability in children with autism.' *Acta Paediatrica* 85(9):1076–1079.

Desser, L., and Rehberger, A. (1990) 'Induction of tumor necrosis factor in human peripheral-blood mononuclear cells by proteolytic enzymes.' *Oncology* 47:475–477.

Desser, L., Rehberger, A., and Paukovits, W. (1994) 'Proteolytic enzymes and amylase induce cytokine production in human peripheral blood mononuclear cells in vitro.' *Cancer Biotherapy* 9:253–263.

Dooley, C.P., Saad, C., and Valenzuela, J.E. (1988) 'Studies of the role of opioids in control of human pancreatic secretion.' *Digestive Diseases and Sciences* 33(5):598–604.

Doumas, A., van den Broek, P., Affolter, M., and Monod, M. (1998) 'Characterization of the prolyl dipeptidyl peptidase gene (dppIV) from the koji mold Aspergillus oryzae.' *Applied Environmental Microbiology* December *64*(12):4809-15.

Duskova, M., and Wald, M. (1999) 'Orally administered proteases in aesthetic surgery.' *Aesthetic Plastic Surgery 23*(1):41–44.

Edelson, S.B., and Cantor, D.S. (1998) 'Autism: Xenobiotic influences.' *Toxicology and Industrial Health 14*(4):553–563.

Edwards, R., Peet, M., Shay, J., and Horrobin, D. (1998) 'Omega-3 polyunsaturated fatty acid levels in the diet and in red blood cell membranes of depressed patients.' *Journal of Affective Disorders 48*(2–3):149–155.

Egger, J., Carter, C.H., Soothill, J.F., and Wilson, J. (1992) 'Effect of diet treatment on enuresis in children with migraine or hyperkinetic behavior.' *Clinical Pediatrics (Phila)* 31(5):302–7.

Egger, J., Carter, C.M., Soothill, J.F., and Wilson, J. (1989) 'Oligoantigenic diet treatment of children with epilepsy and migraine.' *Journal of Pediatrics 114*(1):51–8.

Fehr, K. (1984) 'Die Bedeutung von Immunprozessen in der Pathogenese Entzündlich-Rheumatischer Erkrankungen.' *Aktuelle Rheumatologie 9*:1–12.

Felton, G. (1977) 'Does kinin released by pineapple stem bromelain stimulate production of prostaglandin E1-like compounds?' *Hawaii Medical Journal 2*:39-47.

Fiasse, R., Lurhuma, A.Z., Cambiaso, C.L., Masson, P.L., and Dive, C. (1978) 'Circulating immune complexes and disease activity in Crohn's disease.' *Gut 19*:611–617.

Friess, H., Bohm, J., and Ebert, M. (1993) 'Enzyme treatment after gastrointestinal surgery.' *Digestion 54*(2):48–53.

Friess, H., Kleeff, J., Malfertheiner, P., Homuth, K., and Buchler, M.W. (1998) 'Influence of high-dose pancreatic enzyme treatment on pancreatic function in health volunteers.' *International Journal of Pancreatology 23*(2):115–23.

Gerard, G. (1972) 'Anti-cancer therapy with bromelain.' *Agressologie 13*:261–274.

Glenk, W., and Neu, S. (1990) *Enzyme, Die Bausteine des Lebens: Wie Sie Wirken, Helfen und Heilen.* Munich: Wilhelm Heyne Verlag.

Gobbi, G., Bouquet, F., Greco, L., Lambertini, A., Tassinari, C.A., Ventura, A., and Zaniboni, M.G. (1992) 'Coeliac disease, epilepsy and cerebral calcifications.' *The Lancet 340*:439–443.

Gonzalez, N.J., and Isaacs, L.L. . (1999) 'Evaluation of pancreatic proteolytic enzymes treatment of adenocarcinoma of the pancreas, with nutrition and detoxification support.' *Nutrition and Cancer 33*(2):117–124.

Graham, D.Y. (1977) 'Enzyme replacement therapy of exocrine pancreatic insufficiency in man.' *New England Journal of Medicine 296*:1314–1317.

Griffin, S.M., Alderson, D., and Farndon, J.R. (1989) 'Acid resistant lipase as replacement therapy in chronic pancreatic exocrine insufficiency: A study in dogs.' *Gut 30*:1012–1015.

Gupta, S., Aggarwal, S., Rashanravan, B., and Lee, T. (1998) 'Th1- and Th2-like cytokines in CD4+ and CD8+ T cells in autism.' *Journal of Neuroimmunology 85*(1):106–109.

Gutfreund, A., Taussig, S., and Morris, A. (1978) 'Effect of oral bromelain on blood pressure and heart rate of hypertensive patients.' *Hawaii Medical Journal 37*:143–146.

Gracia, M.I., Aranibar, M.J., Lazaro, R., Medel, P., and Mateos, G.G. (2003) 'Alpha-amylase supplementation of broiler diets based on corn.' *Poultry Science* March 82(3):436-42.

Hadjivassiliou, M., Grunewald, R.A., Lawden, M., Davies-Jones, G.A., Powell, T., and Smith, C.M. (2001) 'Headache and CNS white matter abnormalities associated with gluten sensitivity.' *Neurology* 56(3):385–388.

Hallert, C., Astom, J., and Sedvall, G. (1982) 'Psychic disturbances in adult coeliac disease. III. Reduced central monoamine metabolism and signs of depression.' *Scandinavian Journal of Gastroenterology* 17:25–28.

Halsted, C.H. (1999) 'The relevance of clinical nutrition education and role models to the practice of medicine.' *European Journal of Clinical Nutrition* 53, Suppl 2:S29–34.

Harbige, L.S. (1996) 'Nutrition and immunity with emphasis on infection and autoimmune disease.' *Nutrition and Health* 10(4):285–312.

Harris, R.M., Picton, R., Singh, S., and Waring, R.H. (2000) 'Activity of phenolsulfotransferases in the human gastrointestinal tract.' *Life Sciences* September 15;67(17):2051-7.

Hart, A., Kamm, M.A. (2002) 'Mechanisms of initiation and perpetuation of gut inflammation by stress.' *Alimentary Pharmacology and Therepeutics* December 16(12):2017-28.

Hausch, F., Shan, L., Santiago, N.A., Gray, G.M., and Khosla, C. (2002) 'Intestinal digestive resistance of immunodominant gliadin peptides.' *American Journal of Physiology Gastrointestinal and Liver Physiology* October 283(4):G996-G1003.

Heisler, M. 'Comment: secretin for autism: unproven treatment or ineffective treatment?' *The Annals of Pharmacotherapy* 36(7):1296–1296.

Hekkens, W. (1978) 'Antibodies to gliadin in serum of normals, coeliac patients, and schizophrenics.' *The Biological Basis of Schizophrenia.* (eds) Hemmings and Hemmings, MTP Press. Lancaster.

Hemmings, W.A., and Williams, E.W. (1978) 'Transport of large breakdown products of dietary protein through the gut wall.' *Gut* 19:715-723.

Herlihy, W. (2001) 'Repligen's Goal: Develop Safe and Effective Drugs for Debilitating Pediatric Disorders through Modern Biotechniology.' Presentation at Defeat Autism Now Conference, San Diego, CA.

Hewitt, H., Whittle, S., Lopez, S., Bailey, E., and Weaver, S. (2000) 'Topical use of papaya in chronic skin ulcer therapy in Jamaica.' *West Indian Medical Journal* 49(1):32–33.

Hodgson, H.J.F., Potter, B.J., and Jewell, D.P. (1977) 'Immune complexes in ulcerative colitis and Crohn's disease.' *Clinical and Experimental Immunology* 29:187–196.

Hollander, D. (1999) 'Intestinal permeability, leaky gut, and intestinal disorders.' *Current Gastroenterology Reports* 1(5):410–416.

Hopwood, D.E., Pethick, D.W., and Hampson, D.J. (2002) 'Increasing the viscosity of the intestinal contents stimulates proliferation of enterotoxigenic Escherichia coli and Brachyspira pilosicoli in weaner pigs.' *The British Journal of Nutrition* November 88(5):523-32.

Horrocks, L.A., and Yeo, Y.K. (1999) 'Health benefits of docosahexaenoic acid.' *Pharmacological Research* 40(3):211–225.

Horvath, K., Papadimitriou, J.C., Rabsztyn, A., Drachenberg, C., and Tildon, J.T. (1999) 'Gastrointestinal abnormalities in children with autistic disorders.' *Jornal de Pediatria* 135(5):559–63.

Horvath, K., Stefanatos, G., Sokolski, K.N., Wachtel, R., Nabors, L., and Tildon J.T. (1998) 'Improved social and language skills after secretin administration in patients with autistic spectrum disorders.' *Journal of the Association for Academic Minority Physicians* 9(1):9–15.

Houck, J.C., Chang, C.M., and Klein, G. (1983) 'Isolation of an effective debriding agent from the stems of pineapple plants.' *International Journal of Tissue Reactions* 5:125–134.

Hunter, L.C., O'Hare, A., Herron, W.J., Fisher, L.A., and Jones, G.E. (2003) 'Opioid peptides and dipeptidyl peptidase in autism.' *Developmental Medicine and Child Neurology* February 45(2):121-8.

Ivaniyta, L.I., Ivaniyta, S.O., Kornatskaya, A.G., and Belis, N.I. (1998) 'Systemic enzyme therapy in the treatment of chronic salpingitis and infertility.' Institute of Pediatrics, Obstetrics, and Gynecology, Ukraine *Farmatsevtychnyi Zhurnal* (Kiev) 2:89-92.

Jager, H. (1990) 'Hydrolytic enzymes in the treatment of patients with HIV-infections.' Lecture delivered at the First International Conference on Systemic Enzyme Therapy, Klagenfurt, Austria, September 12.

Jager, H., Popescu, M., and Kaboth, W. (1987) 'Circulating immune complexes in HIV infection.' Paper presented at the 2nd Symposium on Immunobiology In Clinical Oncology and Immune Dysfunctions, Nice, April 4–7.

Jyonouchi, H., Sun, S., and Le, H. (2001) 'Proinflammatory and regulatory cytokine production associated with innate and adaptive immune responses in children with autism spectrum disorders and developmental regression.' *Journal of Neuroimmunology* 120(1–2):170–179.

Kaplan, B.J., McNicol, J., Conte, R.A. and Moghadam, H.K. (1989) 'Dietary replacement in preschool-aged hyperactive boys.' *Pediatrics* 83(1):7–17.

Kawashima, H., Mori, T., Kashiwagi, Y., Takekuma, K., Hoshika, A., and Wakefield, A. (2000) 'Detection and sequencing of measles virus from peripheral [blood] mononuclear cells from patients with inflammatory bowel disease and autism.' *Digestive Diseases and Sciences* 45:4 pp.723–729.

Kaye, B.R. (1989) 'Rheumatologic manifestations of infection with human immunodeficiency virus (HIV).' *Annals of Internal Medicine* 111:158–167.

Keshavarzian, A., Choudhary, S., Holmes, E.W., Yong, S., Banan, A., Jakate, S., and Fields, J.Z. (2001) 'Preventing gut leakiness by oats supplementation ameliorates alcohol-induced liver damage in rats.' *Journal of Pharmacology and Experimental Therapeutics* 299(2):442–448.

Kidd, P.M. (2000) 'Attention deficit/hyperactivity disorder (ADHD) in children: Rationale for its integrative management.' *Alternative Medicine Review* 5(5):402–428.

Klein, G., Schwann, H., and Kullich, W. (1988) 'Enzymtherapie bei chronischer Polyarthrits.' *Natur- und Ganzheitsmedizin* 1:112–116.

Kleine, M.W., Hörterer, H., Dieter, R., and Pabst, H. (1990) 'Therapie der lateralen Sprunggelenksdistorsion mit hydrolytsichen Enzymen.' *Deutsche Zeitschrift für Sportmedizin* 41:435–439.

Knill-Jones, R.P., Pearce, H., and Batten, J. (1970) 'Comparative trial of Nutrizym in chronic pancreatic insufficiency.' *British Medical Journal* 4:21–24.

Konstantareas, M.M., and Homatidis, S. (1987) 'Ear infections in autistic and normal children.' *Journal of Autism and Developmental Disorders* 17(4):585–94.

Kozielec, T., Starobrat-Hermelin, B., and Kotkowiak, L. (1994) 'Deficiency of certain trace elements in children with hyperactivity.' *Psychiatria Polska* 28(3):345–353.

Kozielec, T., and Starobrat-Hermelin, B. (1997) 'Assessment of magnesium levels in children with attention deficit hyperactivity disorder (ADHD).' *Magnesium Research* 10(2):143–148.

Kre, I., Kojecky, Z., Matouskova, I., and Benysek, L. (1980) 'Crohn's disease, serum immunodepressive factors, and circulating immune complexes.' *Bollettino Dell'Istituto Sieroterapico Milanese* 59:619–624.

Lauer, D., Muller, R., Cott, C., Otto, A., Naumann, M., and Birkenmeier, G. (2001) 'Modulation of growth factor binding properties of alpha2-macroglobulin by enzyme therapy.' *Cancer Chemotherapy and Pharmacology* 47Suppl:S4–9.

Lehmann, P.V. *et al.* (1992) 'Spreading of T cell autoimmunity to cryptic determinants of an autoantigen.' *Nature* 358:155-7.

Leibow, C., and Rothman, S.S. (1975) 'Enteropancreatic circulation of digestive enzymes.' *Science* 189 (4201):472–474.

Leipner, J., and Saller, R. (2000) 'Systemic enzyme therapy in oncology: Effect and mode of action.' *Drugs* 59(4):769–80.

Lennard-Jones, J.E. (1983) 'Functional gastrointestinal disorders.' *New England Journal of Medicine* February 24; 308(8):431-5.

Li, P., Zhou, L., Levine, R.A., and Chey, W.Y. (1995) 'Aspirin inhibits secretagogue-stimulated and postprandial pancreatic exocrine secretion in conscious rats.' *Pancreas* January 10(1):85-92.

Linskens, R.K., Huijsdens, X.W., Savelkoul, P.H., Vandenbroucke-Grauls, C.M., and Meuwissen, S.G. (2001) 'The bacterial flora in inflammatory bowel disease: Current insights in pathogenesis and the influence of antibiotics and probiotics.' *Scandinarin Journal of Gastroenterology* Suppl (234) 29–40.

Loehry, C.A., Axon, A.T., Hilton, P.J., Hider, R.C., and Creamer, B. (1970) 'Permeability of the small intestine to substances of different molecular weight.' *Gut* 11:446-470.

Lucarelli, S., Lendvai, D., Frediani, T., Finamore, G., Grossi, R., Barbato, M., Zingoni, A.M., and Cardi, E. (1990) 'Hemicrania and food allergy in children.' *Minerva Pediatrica* 42(6):215–218.

Lucarelli, S., Frediani, T., Zingoni, A.M., Ferruzzi, F., Giardini, O., Quintieri, F., Barbato, M., D'Eufemia, P., and Cardi, E. (1995) 'Food allergy and infantile autism.' *Panminerva Medica* 37(3):137–141.

Maehden, K. (1978) 'Enzyme treatment in diseases of the veins.' *Die Arztpraxis* Volume 2 9:14.

Martineau, J., Barthelemy, C., Garreau, B., and Lelord, G. (1985) 'Vitamin B6, magnesium, and combined B6-Mg: Therapeutic effects in childhood autism.' *Biological Psychiatry* 20(5):467–478.

Martineau, J., Barthelemy, C., and Lelord, G. (1986) 'Long-term effects of combined vitamin B6-magnesium administration in an autistic child.' *Biological Psychiatry* 21(5–6):511–518.

Masson, M. (1995) 'Bromelain in blunt injuries of the locomotor system. A study of observed applications in general practice.' *Fortschritte der Medizin (Munchen)* *113*:303–306.

Mertin, J., and Stauder, G. (1997) 'Use of oral enzymes in multiple sclerosis patients.' *International Journal of Tissue Reactions* XIX (1/2), abstracts of 7th Interscience World Conference on Inflammation, Antirheumatics, Analgesics, Immunomodulators, May 19-21, Geneva, Switzerland.

McCann, M. (1993) 'Pancreatic enzyme supplement for treatment of multiple food allergies.' *Annals of Allergy* 71:269.

McCarthy, C.F. (1976) 'Nutritional defects in patients with malabsorption.' *Proceedings of the Nutrition Society 35*:37–40.

McDonald, D.E., Pethick, D.W., Mullan, B.P., and Hampson, D.J. (2001) 'Increasing viscosity of the intestinal contents alters small intestinal structure and intestinal growth, and stimulates proliferation of enterotoxigenic *Escherichia coli* in newly-weaned pigs.' *The British Journal of Nutrition* October *86*(4):487-98.

McDougle, Naylor, S.T., Cohen, D.J., Aghajanian, G.K., Heninger, G.R., and Price, L.H. (1996) 'Effects of tryptophan depletion in drug-free adults with autistic disorder.' *Archives of General Psychiatry 53*(11):993–1000.

McFadden, S.A. (1996) 'Phenotypic variation in xenobiotic metabolism and adverse environmental response: Focus on sulfur-dependent detoxification pathways.' *Toxicology 111*(1–3):43–65.

Mohan, V., Poongothai, S., and Pitchumoni, C.S. (1998) 'Oral pancreatic enzyme therapy in the control of diabetes mellitus in tropical calculous pancreatitis.' *Internation Journal of Pancreatology 24*(1):19–22.

Molloy, C.A., Manning-Courtney, P., Swayne, S., Bean, J., Brown, J.M., Murray, D.S., Kinsman, A.M., Brasington, M., and Ulrich, C.D. 2nd. (2002) 'Lack of benefit of intravenous synthetic human secretin in the treatment of autism.' *Journal of Autism and Developmental Disorders* December *32*(6):545-51.

Monograph: Bromelain. (1998) *Alternative Medicine Review 3*(4):302–305.

Musca, A., Cordova, C., Barnaba, V., Zaccari, C., Levrero, M., van Dyke, A., and Balsano, F. (1984) 'Circulating HbsAg/IgM Complexes in acute and chronic hepatitis B.' *Hepato-Gastroenterology 31*:208–210.

Nagakura, T., Matsuda S., Shichijyo K., Sugimoto H., and Hata K. (2000) 'Dietary supplementation with fish oil rich in omega-3 polyunsaturated fatty acids in children with bronchial asthma.' *European Respiratory Journal 16*(5):861–865.

Neubauer, R.A. (1961) 'A plant protease for potentiation of and possible replacement of antibiotics.' *Experimental Medicine and Surgery 19*:143–160.

Neuhofer, C. (1990) 'Multiple sclerosis: treatment with enzyme preparations.' First International Conference on Systemic Enzyme Therapy, Klagenfurt, Austria: Mucos Pharma GmbH and Company, September 12.

Nieper, H.A. (1974) 'A program for the treatment of cancer.' *Krebs 6*:124–127.

Nouza, K. (1994) 'Outlooks of systemic enzyme therapy in rheumatoid arthritis and other immunopathological diseases.' *Acta Universitatis Carolinae* (Praha) *40*(1–4):101–4.

Oaten, S. (1993) 'The importance of optimal nutrition in the learning disabled child.' Lecture delivered at the Literacy conference held at the Hornsby International Centre, London, 16–19 September.

Oelgoetz, A.W., Oelgoetz, P.A., and Wittenkind, J. (1935) 'The treatment of food allergy and indigestion of pancreatic origin with pancreatic enzymes.' *American Journal of Digestive Diseases and Nutrition* 2:422–6.

Oderkirk, A. (1996) *Poultry Digest.* Nova Scotia Department of Agriculture and Marketing. April.

O'Reilly, B.A., and Waring, R.H. (1993) 'Enzyme and sulphur oxidation deficiencies in autistic children with known food/chemical intolerances.' *Journal of Orthomolecular Medicine* 8(4):198–200.

Ottokar, R. (1983) 'Adjuvant Enzyme Treatment Before and After Breast Cancer Surgery.' *Deutsche Zeitschrift fur Onkologie 25(5):*130-136.

Owley, T., McMahon, W., Cook, E.H., Laulhere, T., South, M., and Zellmer Mays. L. (2001) 'Multisite, double-blind, placebo-controlled trial of porcine secretin in autism.' *Journal of the American Academy of Child and Adolescent Psychiatry* 40:1293–9.

Pan, B., Li, D., Piao, X., Zhang, L., and Guo, L. (2002) 'Effect of dietary supplementation with alpha-galactosidase preparation and stachyose on growth performance, nutrient digestibility and intestinal bacterial populations of piglets.' *Archiv fur Tierernahrung* October 56(5):327-37.

Phelan, J.J. (1974) 'The nature of gliadin toxicity in celiac disease.' *Biochemical Society Transactions* 2:1368–1370.

Phelan, J.J., Stevens, F.M., and McNicholl, B. (1977) 'Coeliac disease: The abolition of gliadin toxicity by enzymes from *Aspergillus niger.*' *Clinical Science and Molecular Medicine* 53:35–43.

Phelan, J.J., Stevens, F.M., Cleer, W.F., McNicholl, B, McCarthy, C.F., and Fottrell, P.F. (1978) 'The detoxification of gliadin by the enzymic cleavage of a side-chain substituent.' *Perspectives in Coeliac Disease* Eds. B. McNicholl, C.F. McCarthy, and P.F. Fottrell. University Park Press. Baltimore, Maryland.

Posit Health News (1989) Plant enzymes can improve cell-mediated immunity by breaking up circulating immune complexes. Spring *16*:18–19.

Pottinger, F. (1983) 'The Pottinger's Cats Study: A study in nutrition.' *American Journal of Orthodontics and Oral Surgery.* Price-Pottinger Nutritional Foundation 800–862–6759, www.price-pottinger.org.

Prasad, A. (2000) 'Effects of zinc deficiency on immune functions.' *Journal of Trace Elements and Experimental Medicine 13*:1–20.

Ransberger, K., Van Schaik, W., Pollinger, W., and Stauder, G. (1988) 'Naturheilkundliche Therapie von AIDS mit Enzympraparaten.' *Forum des Praktischen und Allgemeinarztes* 4:27.

Ransberger, K., *et al.* (1986) 'Enzyme Therapy in Multiple Sclerosis.' *Der Kassenarzt* 41:42-45.

Reichelt, K.L., Lindback, T., and Scott, H. (1994a) 'Increased levels of antibodies to food proteins in Down syndrome.' *Acta Paediatrica Japonica* 36:489–492.

Reichelt, K.L. and Knivsberg, A.M. (2003) 'Can the pathophysiology of autism be explained by the nature of the discovered urine peptides?' *Nutritional Neuroscience* February 6(1):19-28.

Richardson, A.J., and Ross, M.A. (2000) 'Fatty acid metabolism in neurodevelopmental disorder: A new perspective on associations between attention deficit/hyperactivity disorder, dyslexia, dyspraxia, and the autistic spectrum.' *Prostaglandins, Leukotrienes, and Essential Fatty Acids 63*(1–2):1–9.

Roberts, W., Weaver, L., Brian, J., Bryson, S., Emelianova, S., Griffiths, A.M., MacKinnon, B., Yim, C., Wolpin, J., and Koren, G. (2001) 'Repeated doses of porcine secretin in the treatment of autism: a randomized, placebo-controlled trial.' *Pediatrics* May *107*(5):E71.

Rogers, A.E., Zeisel, S.H., and Groopman, J. (1993) 'Diet and carcinogenesis.' *Carcinogenesis 14*(11):2205–2217.

Rolfe, R.D. (2000) 'The role of probiotic cultures in the control of gastrointestinal health.' *Journal of Nutrition* 130(2S Suppl):396S–402S. [Includes a summary and review of 85 studies.]

Rothman, S., Liebow, C., and Isenman, L. (2002) 'Conservation of digestive enzymes.' *Physiological Reviews 82*(1):1–18.

Sakalova, A., Mikulecky, M., Holomanova, D., Langner, D., Ransberger, K., Stauder, G., Mistrik, M., Gazova, S., Chabronova, I., Benzova, M. (1992) 'The favorable effect of hydrolytic enzymes in the treatment of immunocytomas and plasmacytomas.' *Vnitrni Lekarstvi* September *38*(9):921-9.

Salerno, G., De Franco, A., La Rosa, S., and Calistro, V. (2002) 'Malabsorption syndromes.' *Rays* January-March *27*(1):19-34.

Sandler, R.H., Finegold, S.M., Bolte, E.R., Buchanan, C.P., Maxwell, A.P., Vaisanen, M.L., Nelson, M.N., and Wexler, H.M. (2000) 'Short-term benefit from oral vancomycin treatment of regressive-onset autism.' *Journal of Child Neurology 15*(7):429–435.

Sansum (1932) 'The Treatment of Indigestion, Underweight, and Allergy' *Southwestern Medicine 16*:452–462.

Sarosiek, J., Slomiany, A., and Slomiany, B.L. (1988) 'Evidence for weakening of gastric mucus integrity by *Campylobacter pylori*.' *Scandinavian Journal of Gastroenterology* June *23*(5):585-90.

Sastry, K.V., and Gupta, P.K. (1978) 'In vitro inhibition of digestive enzymes by heavy metals and their reversal by chelating agent: Part I. Mercuric chloride intoxication.' *Bulletin of Environmental Contaminants and Toxicology* December *20*(6):729-35. [Part II same reference: *20*(6):736-42]

Scheef, W. (1987) 'Enzymtherapie, Lehrbuch der Naturheilverfahren.' *Hippokrates-Verlag* Bd. II, S. 95–103. (Hrgs. K. –Ch. Schimmer)

Schneider, M.U., Knoll-Ruzicka, M.L., Domschke, S., Heptner, G., and Domschke, W. (1985) 'Pancreatic enzyme replacement therapy: Comparative effects of conventional and enteric-coated microspheric pancreatin and acid-stable fungal enzyme preparations on steatorrhea in chronic pancreatic.' *Hepato-Gastroenterology 32*:97–102.

Schwimmer, S. (1981) *Source Book of Food Enzymology*. The AVI Publishing Company. Westport, Connecticut.

Seim, A.R., and Reichelt, K.L. (1995) 'An enzyme/brain-barrier theory of psychiatric pathogenesis: Unifying observations on phenylketonuria, autism, schizophrenia, and postpartum psychosis.' *Medical Hypotheses 45*(5):498–502.

Seligman, B. (1962) 'Bromelain: An Anti-inflammatory Agent.' *Angiology 13*:508–510.

Singer, F. (1990) 'Aktivierte Arthrosen Knorpelschonend Behandeln.' Lecture delivered at the Systemische Enzyme-therapie 10th Symposium, Frankfurt, organized by Medizinische Enzyme-Forschungsgesellschaft e.V.

Singh, V.K. (1998a) 'Association of anti-MBP and anti-NAFP antibodies with HHV-6 antibodies in a child with autistic regression.' *Journal of Allergy and Clinical Immunology 101:*1(part 2) S122.

Singh, V.K. (1998b) 'Serological association of measles virus and human herpes virus-6 with brain autoantibodies in autism.' *Clinical Immunology and Immunopathology 89(* 1):105–108.

Singh, V., and Nelson, C. (2002) 'Abnormal Measles Serology and Autoimmunity in Autistic Children' *Journal of Allergy and Clinical Immunology* January *109*(1):S232,

Singh, V.K., Lin, S.X., Newell, E., and Nelson, C. (2002) 'Abnormal measles-mumps-rubella antibodies and CNS autoimmunity in children with autism.' *Journal of Biomedical Science* July-August *9*(4):359-64

Sponheim, E., Oftedal, G., and Helverschou, S.B. (2002) 'Multiple doses of secretin in the treatment of autism: a controlled study.' *Acta Paediatria 9 1*(5):540-5

Starley, I.F., Mohammed, P., Schneider, G., and Bickler, S.W. (1999) 'The treatment of paediatric burns using topical papaya.' *Burns 25*(7):636–639.

Starobrat-Hermelin, B., and Kozielec, T. (1997) 'The effects of magnesium physiological supplementation on hyperactivity in children with attention deficit hyperactivity disorder (ADHD). Positive response to magnesium oral loading test.' *Magnesium Research 10*(2):149–156.

Starobrat-Hermelin, B. (1998) 'The effect of deficiency of selected bioelements on hyperactivity in children with certain specified mental disorders.' *Annales Academiae Medicae Stetinensis 44*:297–314.

Stauder, G. (1990) 'Hydrolytische Enzyme als Adjuvante Therapie bei HIV Infektionen.' *Münchener AIDS-Tage: Abstraktband S,* 36–37.

Stauder, G. (1995) 'Pharmacological effects of oral enzyme combinations.' *Casopis Lekaru Ceskych 134*(19):620–624.

Stauder, G., Fuchs, D., Jager, H., Samtleben, W., and Wachter, H. (1989) 'Adjuvant therapy of HIV infections with hydrolytic enzymes: Course of heopterin, CD4-T-cells, immune complexes (IC), and clinical efficacy.' 8th International Workshop on Biochemical and Clinical Aspects of Peridines, St. Christoph, Austria, February 11–18.

Stauder, G., Ransberger, K., Streichhan, P., Van Schaik, W., and Pollinger, W. (1988) 'The use of hydrolytic enzymes as adjuvant therapy in AIDS/ARC/LAS patients.' *Biomedicine and Pharmacotherapy 42*(1):31–34.

Steffen, C., and Menzel, J. (1983) 'Enzymbbau von Immunkomplexen.' *Zeitschrift für Rheumatologie 42*:249–255.

Steffen, C., Smolen, J., Miehlke, K., Hörger, I., and Menzel, J. (1985) 'Enzyme therapy in comparison to the immune complex determination in chronic polyarthritis.' *Zeitschrift für Rheumatologie 44*:51-56

Stevens, F.M., Phelan, J.J., McNicholl, B., Comerford, F.R., Fottrell, P.F., and McCarthy, C.F. (1978) 'Clinical demonstration of the reduction of gliadin toxicity by enzymic cleavage of side-chain constituent.' *Perspectives in Coeliac Disease* Eds. B. McNicholl, C.F. McCarthy, and P.F. Fottrell. University Park Press. Baltimore, Maryland.

Stoll, A. L. (2001) *The Omega-3 Connection.* New York: Simon & Schuster.

Sturniolo, G.C., Di Leo, V., Ferronato, A., D'Odorico, A., and D'Inca, R. (2001)

'Zinc supplementation tightens 'leaky gut' in Crohn's disease.' *Inflammatory Bowel Diseases* 7(2):94–98.

Suarez, F., Levitt, M.D., Adshead, J., Barkin, and J.S. (1999) 'Pancreatic supplements reduce symptomatic response of healthy subjects to a high fat meal.' *Digestive Diseases and Sciences* 44:1317–21.

Swanson, J., and Kinsbourne, M. (1980) 'Food dyes impair performance of hyperactive children on a laboratory learning test.' *Science Magazine*, March 28 207:1485–1487.

Targoni, O.S., Tary-Tehmann, M., and Lehmann, P.V. (1999) 'Prevention of murine EAE by oral hydrolytic enzyme treatment.' *Journal of Autoimmunity* 12(3):191–8.

Tassman, G.C., Zafran, J.N., and Zayon, G.M. (1964) 'Evaluation of a plant proteolytic enzyme for the control of inflammation and pain.' *Journal of Dental Medicine* 19:73–77.

Taussig, S.J., and Nieper, H.A. (1979) 'Bromelain: Its use in prevention and treatment of cardiovascular disease, present status.' *Journal of the International Academy for Preventive Medicine* 6:139–151.

Taussig, S.J., and Batkin, S. (1988) 'Bromelain, the enzyme complex of pineapple (*Ananas comosus*) and its clinical application. An update.' *Journal of Ethnopharmacology* 22:191-203.

Taussig, S.J., Szekerczes, J., and Batkin, S. (1985) 'Inhibition of tumor growth in vitro by bromelain, an extract of the pineapple plant (*Ananas comosus*).' *Planta Medica* 6:538–539.

Theofilopoulos, A.N. (1980) 'Evaluation and clinical significance of circulating immune complexes.' Progress in Clinical Immunology 4:63–106.

Tinozzi, S., and Venegoni, A. (1978) 'Effect of bromelain on serum and tissue levels of amoxicillin.' Drugs under Experimental and Clinical Research 4:39–44.

Truss, C.O. (1984) 'Metabolic abnormalities in patients with chronic candidiasis.' *Journal of Orthomolecular Psychiatry* 13:66–93.

Ufflemann, K. (1990) 'Enzyme treatment of soft tissue rheumatism.' Lecture delivered at the First International Conference on Systemic Enzyme Therapy, Klagenfurt, Austria, September 12.

Uhlmann, V., Martin, C.M., Sheils, O., Pilkington, L., Silva, I., Killalea, A., Murch, S.B., Wakefield A.J., and O'Leary, J.J. (2002) 'Potential viral pathogenic mechanism for new variant inflammatory bowel disease.' *Journal of Clinical Pathology: Molecular Pathology* 55:0–6.

Vancassel, S., Durand, G., Barthelemy, C., Lejeune, B., Marineau, J., Guilloteau, D., Andres, C., and Chalon S. (2001) 'Plasma fatty acid levels in autistic children.' *Prostaglandins, Leukotrienes, and Essential Fatty Acids* 65(1):1–7.

van Gent, T., Heijnen, C.J., and Treffers, P.D. (1997) 'Autism and the immune system.' *Journal of Child Psychology and Psychiatry* 38(3):337–349.

Veldoza, Z., Martinkova, R., and Maderova, S. (1999) 'Use of Phlogenzym in the treatment of secretory otitis in the outpatient practice.' Congress of the Czech Society of Otorhinolaryngology and Surgery of Head and Neck, September 9-11, Hradec Králové, Czech Republic.

Vogelsang, H., Schwarzenhofer, M., and Oberhuber, G. (1998) 'Changes in gastrointestinal permeability in celiac disease.' *Digestive Diseases* 16(6):333–336.

Wakefield, A.J., Anthony, A., Murch, S.H., Thomson, M., Montgomery, S.M., Davies, S., O'Leary, J.J., Berelowitz, M., and Walker-Smith, J.A. (2000) 'Enterocolitis in children with developmental disorders.' *American Journal of Gastroenterology* 95(9):2285–2295.

Wakefield, A.J., Murch, S.H., Anthony, A., Linnell, J., Casson, D.M., Malik, M., Berelowitz, M., Dhillon, A.P., Thomson, M.A., Harvey, P., Valentine, A., Davies, S.E., Walker-Smith, J.A. (1998) 'Ileal-lymphoid-nodular hyperplasia, non-specific colitis, and pervasive developmental disorder in children.' *The Lancet* 351:637–41.

Walker, W.A. (1975) 'Antigen absorption from the small intestine and gastrointestinal disease.' *Pediatric Clinics of North America* 22:731-746.

Walsh, W., and Anjum, U. (2001) 'Metal-metabolism and autism.' Presentation at the American Psychiatric Association Annual Meeting, May 10, New Orleans.

Waring, R.H., and Ngong, J.M. (1993) 'Sulphate metabolism in allergy-induced autism: Relevance to the disease aetiology.' Conference papers from Biological Perspectives in Autism held at the University of Durham, April. Published by Autism Research Unit, University of Sunderland, United Kingdom.

Warshaw, A.L., Walker, W.A., and Isselbacher, K.J. (1974) 'Protein uptake by the intestine: Evidence for absorption of intact macromolecule's.' *Gastroenterology* 66:987-992.

Wecker, L., Miller, S.B., Cochran, S.R., Dugger, D.L., and Johnson, W.D. (1985) 'Trace element concentrations in hair from autistic children.' *Journal of Mental Deficiency Research* 29(part 1):15-22.

Wendorff, J., Kamer, B., Zielinska, W., and Hofman, O. (1999) 'Allergy effect on migraine course in older children and adolescents.' *Neurologia i Neurochirurgia Polska* 33 Suppl 5:55–65.

White, R.R., Crawley, F.E., Vellini, M., and Rovati, L.A. (1988) 'Bioavailability of 125I bromelain after oral administration to rats.' *Biopharmaceutics and Drug Disposition* 9:397–403.

Whiteley, P and Shattock, P. (2002) 'Biochemical aspects in autism spectrum disorders: updating the opioid-excess theory and presenting new opportunities for biomedical intervention.' *Expert Opinion on Therapeutic Targets* April 6(2):175-83.

Wimalawansa, S.J. (1981) 'Papaya in the treatment of chronic infected ulcers.' *Ceylon Medical Journal* 26(3):129–132.

Wolf, J.L., Rubin, D.H., Finberg, R., Kauffman, R.S., Sharpe A.H., Trier, J.S., and Fields, B.N. (1981) 'Intestinal M cells: A pathway for entry of retrovirus into the host.' *Science* 212:471-472.

Worschhauser, S. (1990) 'Konservative Therapie der Sportverletungen. Enzympraparate für Therapie und Prophylaxe.' *Algemeinmedizin* 19:173.

Wrba, H. (1990) 'New approaches in treatment of cancer with enzymes.' Lecture delivered at the First International Conference on Systemic Enzyme Therapy, September 12.

Yoneyama, H., Suzuki, M., Fujii, K., and Odajima, Y. (2000) 'The effect of DPT and BCG vaccinations on atopic disorders.' *Arerugi* 49(7):585–592.

Zanjanian, M.H. (1976) 'The intestine in allergic diseases.' *Annals of Allergy* September 37(3):208-18.

Further References

Books and references on enzymes and health

www.autism-society.org Autism Society of America

www.enzymedica.com Enzymedica company

www.enzymestuff.com Enzyme information

www.enzymeuniversity.com Enzyme information.

www.houstonni.com Houston Nutraceuticals, Incorporated 1-866-757-8627

www.throppsnutrition.com Thropps Nutrition 1-800-490-9067

Cichoke, A. (1994) *Enzymes and Enzyme Therapy: How to Jump-Start Your Way to Lifelong Good Health.* Keats Publishing. Los Angeles, California.

Crook W. (1987) *Solving the Puzzle of Your Hard-to-Raise Child.* Random House. New York, New York.

Fuller, D. (1998) *The Healing Power of Enzymes.* Forbes Custom Publishing, Thomson Learning. Stamford, Connecticut.

Gershon, M. (1998) *The Second Brain : The Scientific Basis of Gut Instinct and a Groundbreaking New Understanding of Nervous Disorders of the Stomach and Intestines.* HarperCollins Publishers Inc. New York, New York.

Gilbere, G. (2000) *I Was Poisoned By My Body: The Odyssey of a Doctor Who Reversed Fibromyalgia, Leaky Gut Syndrome, and Multiple Chemical Sensitivity – Naturally!* Lucky Press. Lancaster, Ohio.

Gottschall, E. (1994) *Breaking the Vicious Cycle: Intestinal Health Through Diet.* Kirkton Press. Ontario, Canada.

Guyton, A., and Hall, J. *Textbook of Medical Physiology.* (10th edition). W.B. Saunders Company. Philadelphia, Pennsylvania.

Hersey, J. (1996) *Why Can't My Child Behave?* Pear Tree Press, Inc. Alexandria, Virginia.

Howell, E. (1985) *Enzyme Nutrition: The Food Enzyme Concept.* Avery Publishing. Wayne, New Jersey.

Howell, E. (1994) *Food Enzymes for Health and Longevity.* Lotus Press. Twin Lakes, Wisconsin.

In Salinas, A.F., and Hanna, G. M. *Immune Complexes and Human Cancer,* Vol 15. Plenum Press. New York, New York. (eds)

King, J.E. (2000) *Mayo Clinic on Digestive Health.* Mayo Foundation for Medical Education and Research. Rochester, Minnesota.

Kranowitz, C.S. (1988) *The Out-Of-Sync Child: Recognizing and Coping with Sensory Integration Dysfunction.* The Berkeley Publishing Group. New York, New York.

Kranowitz, C.S. (2003) *The Out-Of-Sync Child Has Fun: Activities for Kids with Sensory Integration Dysfunction.* Perigee. New York, New York.

Lee, L., Turner, L., and Goldberg, B. (1998) *The Enzyme Cure: How Plant Enzymes Can Help You Relieve Health Problems.* Future Medicine Publishing. Tiburon, California.

Lipski, E. (1996) *Digestive Wellness.* Keats Publishing. Los Angeles, California.

Loomis, H.F. Jr., and Davis, A. (1999) *Enzymes: The Key to Health.* Grote Publishing. Madison, Wisconsin.

Lopez, D.A., Williams, M., and Miehlke. (1994) *Enzymes – The Fountain of Life.* The Neville Press. Charleston, South Carolina.

Martin, J., and Rona, Z. (1996) *The Complete Candida Yeast Guidebook.* Prima Publishing. Roseville, California.

Murray, M.T. (1997) *Chronic Candidiasis: Your Natural Guide to Healing with Diet, Vitamins, Minerals, Herbs, Exercise, and Other Natural Methods.* Prima Publishing. Roseville, California.

Nichols, T.W., and Faas N. (1999) *Optimal Digestion: New Strategies for Achieving Digestive Health.* HarperCollins Publishers Inc. New York, New York.

Rapp, D. (1991) *Is This Your Child?* William Morrow and Company. New York, New York.

Rochlitz, S. (2000) *Allergies and Candida.* Human Ecology Balancing Science.

Santillo, H. (1993) *Food Enzymes: The Missing Link to Radiant Health.* Prescott, Arizona. Hohm Press.

Sears, B. (2002) *The Omega Rx Zone: The Miracle of the New High-Dose Fish Oil.* HarperCollins/Regan Books. New York, New York.

Shaw, W. (2001) *Biological Treatments for Autism and PDD.* Great Plains Laboratory Inc. Lenexa, Kansas.

Thompson, W.G. (1989) *Gut Reactions: Understanding Symptoms of the Digestive Tract.* Plenum Press. New York, New York.

Truss, C. Orian. (1983) *Missing Diagnosis.* Missing Diagnosis Publishing. Birmingham, Alabama.

Wolf, M., and Ransberger, K. (1972) *Enzyme Therapy.* New York. Vantage Press. out of print.

There are many good books available on food additives, food and chemical sensitivities/intolerances/allergies, nutrition, and effects on behavior as well.

Even Further References

Sensory integration, migraines

The Sensory Integration Resource Center (*www.sinetwork.org*)

Mansfield, L.E., Vaughan, T.R., Waller, S.F., Haverly, R.W., and Ting, S. (1985) 'Food allergy and adult migraine: Double-blind and mediator confirmation of an allergic etiology.' *Annals of Allergy* 55(2):126–129.

Ratner, D., Shoshani, E., and Dubnov, B. (1983) 'Milk protein-free diet for nonseasonal asthma and migraine in lactase-deficient patients.' *Israel Journal of Medical Sciences* 19(9):806–809.

Trotsky, M.B. (1994) 'Neurogenic vascular headaches, food, and chemical triggers.' *Ear Nose Throat Journal* 73(4):228–230, 235–236.

Vaughan, T.R. (1994) 'The role of food in the pathogenesis of migraine headache.' *Clinical Reviews in Allergy* Summer 12(2):167–180.

Food intolerances/Sensitivities/Allergies

Amiche, M., Delfour, A., and Nicolas, P. (1988) 'Structural requirements for dermorphin opioid receptor binding.' *International Journal of Peptide and Protein Research* 32:1 28–34.

Cooke, W.T., and Smith W.T. (1966) 'Neurological disorders associated with adult coeliac disease.' *Brain* 89:683–722.

Finelli, P.F., McEntee, W.J., Ambler, M., and Kestenbaum, D. (1980) 'Adult coeliac disease presenting as cerebellar syndrome.' *Neurology* 30:245–249.

Gardner, M.L.G. (1994) 'Absorption of intact proteins and peptides.' In L.R. Johnson (ed.) *Physiology of the Gastrointestinal Tract* (3rd edition) New York: Raven Press.

Gardner, M.L.G., and Steffens, K-J (1995) *Absorption of orally administered enzymes*. Berlin and Heidelberg: Springer-Verlag. (eds)

Husby, S., Jensenius, J.C., and Svehag, S.E. (1985) 'Passage of undegraded dietary antigen into the blood of healthy adults. Quantification, estimation of size distribution, and relation of uptake to levels of specific antibodies.' *Scandinavian Journal of Immunology* 22:83–92.

Improta, G., Broccardo, M., Lisi, A., and Melchiorri, P. (1982) 'Neural regulation of gastric acid secretion in rats: Influence of dermorphin.' *Regulatory Peptides* 3:3–4, 251–256.

Kinney, H.C., Burger, P.C., Hurwitz, B.J., Hijmans, J.C., and Grant, J.P. (1982) 'Degeneration of the central nervous system associated with coeliac disease.' *Journal of the Neurological Sciences* January 53(1):9–22.

Karjalainen, J., Martin, J.M., Knip, M., Ilonen, J., Robinson, B.H., Savilahti, E., Akerblom, H.K., and Dosch, H.M. (1992) 'Bovine albumin peptide as a possible trigger of insulin-dependent diabetes mellitus.' *New England Journal of Medicine* *327*:302–307.

Leboyer, M., Bouvard, M.P., Launay, J.M., Recasens, C., Plumet, M.H., Waller-Perotte, D., Tabuteau, F., Bondoux, D., and Dugas, M. (1993) 'Opiate hypothesis in infantile autism? Therapeutic trials with naltrexone.' *Encephale 19*(2):95–102.

Reichelt, W.H., and Reichelt, K.L. (1997) 'The possible role of peptides derived from food proteins in diseases of the nervous system.' *Epilepsy and other neurological disorders in coeliac disease.* John Libbey and Company pp 227-237.

Reichelt, K.L., *et al.* (1990) 'The effect of gluten-free diet on glycoprotein attached urinary peptide excretion and behaviour in schizophrenics.' *Journal Orthomology Medicine* 5:223–239.

Reichelt, K.L., Knivsberg, A.M., Nodland, M., and Lind G. (1994b) 'Nature and consequences of hyperpeptiduria and bovine casomorphin found in autistic syndromes.' *Developmental Brain Dysfunction* 7:71–85.

Shattock, P., Kennedy, A., Rowell, F., and Berney, T. (1990) 'Role of neuropeptides in autism and their relationships with classical neurotransmitters.' *Brain Dysfunction 3*:328–346.

Singh, M.M., and Kay, S.R. (1976) 'Wheat gluten as a pathogenic factor in schizophrenia.' *Science 191*:401–402.

Zagon, I.S., and McLaughlin, P.J. (1987) 'Endogenous opioid systems regulate m-cell proliferation in the developing rat brain.' *Brain Research 412*:68–72.

Zagon, I.S., and McLaughlin, P.J. (1991) 'Identification of opioid peptides regulating proliferation of neurones and glia in the developing nervous system.' *Brain Research 542*:318–332.

Enzymes and disease

Ambrus, J.L., Lassman, H.B., and DeMarchi, J.J. (1967) 'Absorption of exogenous and endogenous proteolytic enzymes.' *Clinical Pharmacology and Therapeutics 8*:362–8.

Bergkvist, R., and Svard, P.C. (1964) 'Studies on the thrombolytic effect of a protease from *Aspergillus oryzae.*' *Acta Physiologica Scandinavica 60*:363-371.

Bonney, R.J. and Davies, P. (1984) 'Possible autoregulatory functions of the secretory products of mononuclear phagocytes.' In D.O. Adams, and G.M. Hanna (eds) *Macrophage Activation*, Vol 13. Plenum Press. New York, New York.

Castell, J.V., Friedrich, G., Kuhn, C.S., and Poppe, G.E. (1997) 'Intestinal absorption of undegraded proteins in men: presence of bromelain in plasma after oral intake.' *American Journal of Physiology* July *273*(1 Pt 1):G139-G146.

Cichoke, A.J. (1995) 'The effect of systemic enzyme therapy on cancer cells and the immune system.' Townsend Letter for Doctors and Patients November 30–2 [review].

Deitrick, R.E. (1965) 'Oral proteolytic enzymes in the treatment of athletic injuries: a double-blind study.' *Pennsylvania Medical Journal* October 35–37.

Fitzgerald, D.E., and Frisch, E.P. (1973) 'Relief of chronic peripheral artery obstruction by intravenous brinase.' *Irish Medical Association* 66:3.

Fitzgerald, D.E., Frisch, E.P., and Milliken, J.C. (1979) 'Relief of chronic arterial obstruction using intravenous brinase.' *Scandinavian journal of Thoracic and Cardiovascular Surgery 13*:327-332.

Frisch, E.P., and Blomback, M. (1979) 'Blood coagulation studies in patients treated with brinase.' In: Progress in Chemical Fibrinolysis and Thrombolysis. Vol. IV, J.F. Davidson (Ed.), Edinburgh: Churchill-Livingstone, pp 184-87.

Gonzalez, N.J., and Isaacs, L.L. (1999) 'Evaluation of pancreatic proteolytic enzyme treatment of adenocarcinoma of the pancreas, with nutrition and detoxification support.' *Nutrition and Cancer 33*:117–24.

Griffin, S.M., Alderson, D., and Farndon, J.R. (1989) 'Acid resistant lipase as replacement therapy in chronic exocrine insufficiency: a study in dogs.' *Gut 30*:1012-1015.

Heinicke, R.M., Van der Wal, M., and Yokoyama, M.M. (1972) 'Effect of bromelain (ananase) on human platelet aggregation.' *Experientia 28*:844–845.

Hellstrom, K.E., Hellstrom, I., Snyder, H.W. Jr., Balint, J.P., and Jones, F.R. (1985) 'Blocking (suppressor) factors, immune complexes and 'extracorporeal immunoadsorption in tumor immunity.'

Hunter, R.G., Henry, G.W., and Heinicke, R.M. (1957) 'The action of papain and bromelain on the uterus.' *American Journal of Obstetrics and Gynecology 73*:867–873.

Kidd, P.M. (2001) 'Multiple sclerosis, an autoimmune inflammatory disease: Prospects for its integrative management.' *Alternative Medicine Reveiw 6*(6):540–566.

Kiesslling, H., and Svenson, R. (1970) 'Influence of an enzyme from *Aspergillus oryzae*, Protease 1, on some components of the fibrinolytic system.' *Acta Chemica Scandinavica 24*:569-579.

Kumakura, S., Yamashita, M., and Tsurufuji, S. (1988) 'Effect of bromelain on kaolin-induced inflammation in rats.' *European Journal of Pharmacology 150*:295–301.

Laidet, B., and Letourneur, M. (1993) 'Enzymatic debridement of leg ulcers using papain.' *Annales de Dermatologie et de Venereologie 120*(3):248.

Larson, L.J., et al. (1988) 'Properties of the complex between alpha-2-macro-globulin and brinase, a proteinase from *Aspergillus oryzae* with thrombolytic effect, thrombosis.' *Research 49*:55-68.

Layer, P., and Groger, G. (1993) 'Fate of pancreatic enzymes in the human intestinal lumen in health and pancreatic insufficiency.' *Digestion 54*(suppl 2):10–4.

Lund, F., Ekestrom, S., Frisch, E.P. and Magaard, F. (1975) 'Thrombolytic treatment with i.v. brinase in advance arterial obliterative disease.' *Angiology 26*:534.

Milla, C.E., Wielinski, C.L., and Warwick, W.J. (1994) 'High-strength pancreatic enzymes.' *Lancet 343*:599 [letter].

Muller-Hepburn, W. (1970) 'Anwendung von Enzymen in der Sportmedizin.' *Forum des Praktischen Arztes* 18.

Nakamura, T., Tandoh, Y., Terada, A., Yamada, N., Watanabe, T., Kaji, A., Imamura, K., Kikuchi, H., and Suda, T. (1998) 'Effects of high-lipase pancreatin on fecal fat, neutral sterol, bile acid, and short-chain fatty acid excretion in patients with pancreatic insufficiency resulting from chronic pancreatitis.' *International Journal of Pancreatology 23*:63–70.

Nieper, H.A. (1977) 'Decrease of the incidence of coronary heart infarct by Mg- and K-orotate and bromelain.' *Acta Medica Empirica 12*:614–618.

Nieper, H.A. (1978) 'Effect of bromelain on coronary heart disease and angina pectoris.' *Acta Medica Empirica 5*:274–278.

Nydegger, U.E. (1979) 'Biological properties and detection of immune complexes in animal and human pathology.' *Reviews of Physiology Biochemistry and Pharmacology 85*:64–111.

Samtleben W., and Gurland, H.J. (1982) 'Plasmapherese bei Lupusnephritis: Rationale Basis und Klinische Erfahrungen.' *Nieren- und Hochdruckkrankheiten 3*:104–108.

Tassman, G.C., Zafran, J.N., and Zayon, G.M. (1965) 'A double-blind crossover study of a plant proteolytic enzyme in oral surgery.' *Journal of Dental Medicine 20*:51–54.

Vanhove, P., Donati, M.B., Claeys, H., Verhaeghe, R., and Vermylen, J. (1979) 'Action of brinase on human fibrinogen and plasminogen.' *Thrombos Haemostas 42*:571-581.

Walker, J.A., Cerny, F.J., Cotter, J.R., and Burton, H.W. (1992) 'Attenuation of contraction-induced skeletal muscle injury by bromelain.' *Medicine and Science in Sports and Exercise* January *24*(1):20-25.

Intestinal impermeability – Leaky gut

Andre, C., Lambert, R., Bazin, H., and Heremans, J.F. (1974) 'Interference of oral immunization with the intestinal absorption of heterologous albumin.' *European Journal of Immunology 4*:701-704.

Avakian, S. (1964) 'Further studies on the absorption of chymotrypsin.' *Clinical Pharmacology and Therapeutics 5*:712–5.

Bjarnason, I., et al. (1984) 'Intestinal permeability in celiac sprue, dermatitis herpetiformis, schizophrenia and atopic eczema.' *Gastroenterology 86*:1029.

Bockman, D.E. and Winborm, W.B. (1966) 'Light and electron microscopy of intestinal ferritin absorption: Observations in sensitized and non-sensitized hamsters.' *The Anatomical Record 155*:603-622.

Fratkin, J.P. (1999) *Leaky Gut Syndrome: A Modern Epidemic.* www.gsdl.com/ news/1999/19990227/index.html.

Gardner, M.L.G. (1988) 'Gastrointestinal absorption of intact proteins.' *Ann. Rev. Nutr. 8*:329-350.

Husby, S., et al. (1987) 'Passage of dietary antigens into the blood of children with coeliac disease: Quantification and size distribution of absorbed antigens.' *Gut 28*:1062-1072.

Husby, S., Jensenius, J.C., and Svehag, S.E. (1986) 'Passage of undergrade dietary antigen into the blood of healthy adults: further characterization of the kinetics of uptake and the size distribution of the antigen.' *Scandanavian Journal of Immunology 24*:447-455.

Jackson, P.G., Lessof, M.H., Baker, R.W. Ferrett, J., MacDonald, D.M. (1981) 'Intestinal permeability in patients with eczema and food allergy. *Lancet 1*:1285-1286.

Johnson S. (2001) 'Micronutrient accumulation and depletion in schizophrenia, epilepsy, autism and Parkinson's disease?' *Medical Hypotheses* May 56(5):641-5.

Lezak, M. (1999) *Inflammatory Conditions and the Gastrointestinal Tract:* www.gsdl.com/news/1999/19990228/index.html.

Menzies, I.S. (1984) 'Transmucosal passage of inert molecules in health and disease.' In Intestinal Absorption and Secretion, E. Skadhauge and K. Heintze, eds., MTP Press, Lancaster, pp 527-543.

Udall, J.N., and Walker, W.A. (1982) 'The physiologic and pathologic basis for the transport of macromolecule's across the intestinal tract.' *Journal of Pediatrics Gastroenterology and Nutrition* 1(3):295-301.

Nutrient deficiencies and malabsorption

Coniglio, S.J., Lewis, J.D., Lang, C., Burns, T.G., Subbani-Siddique, R., and Weintraub, A. (2001) 'A randomized, double-blind, placebo-controlled trial of single-dose intravenous secretin as treatment for children with autism.' *Journal of Pediatrics* 138(5):649–55.

Corbett, B., Khan, K., Czapansky-Beilman, D., Brady, N., Dropik, P., Goldman, D.Z., Delaney, K., Sharp, H., Mueller, I., Shapiro, E., and Ziegler, R. (2001) 'A double-blind, placebo-controlled crossover study investigating the effect of por-cine secretin in children with autism.' *Clinical Pediatrics* (Phila) June 40(6):327-31.

Dannaeus A., et al. (1979) 'Intestinal uptake of ovalbumin in malabsorption and food allergy in relation to serum IgG antibody and orally administrated sodium chromoglycate.' *Clinical Allergy* 9:263-270.

DiMagno, E.P., Go V.L., and Summerskill, W.H. (1973) 'Relations between pancreatic enzyme outputs and malabsorption in severe pancreatic insufficiency.' *New England Journal of Medicine* 228:813-815.

Dunn-Geier, J., Ho, H.H., Auersperg, E., Doyle, D., Eaves, L., Matsuba, C., Orrbine, E., Pham, B., and Whiting, S. (2000) 'Effect of secretin on children with autism: a randomized controlled trial.' *Developmental Medicine and Child Neurology* December 42(12):796-802.

Heatley, R.V. (1986) 'Assessing Nutritional State in Inflammatory Bowel Disease.' *Gut* November 27 (supp) 1:61-6.

Kubena, K.S., McMurray, D.N. (1996) 'Nutrition and the immune system: a review of nutrient-nutrient interactions.' Journal of the American Dietetic Association November 96(11):1156-64.

McCarthy, C.F. (1976) 'Nutritional defects in patients with malabsorption.' *Proceedings of the Nutrition Society* 35:37-40.

Quigley, E.M., and Hurley D. (2000) 'Autism and the gastrointestinal tract.' *American Journal of Gastroenterology* 95:2154–2156.

Unis, A.S., Munson, J.A., Rogers, S.J., Goldson, E., Osterling, J., Gabriels, R., Abbott, R.D., and Dawson, G. (2002) 'A randomized, double-blind, placebo-controlled trial of porcine versus synthetic secretin for reducing symptoms of autism.' *Journal of the American Academy of Child and Adolescent Psychiatry* November 41(11):1315-21.

Life in the gut – Dysbiosis

Carroccio, A., Guarino, A., Zuin, G., Verghi, F., Berni Canani, R., Fontana, M., Bruzzese, E., Montalto, G., and Notarbartolo, A. (2001) 'Efficacy of oral pancreatic enzyme therapy for the treatment of fat malabsorption in HIV-infected patients.' *Alimentary Pharmacology and Therapeutics* 15(10):1619–1625.

Kabil, S., and Stauder, G. (1997) 'Oral enzyme therapy in hepatitis C patients.' *International Journal of Tissue Reactions* 1-2.

Kiprov, D.D., and Lippert, R. (1986) 'The use of plasmapheresis, lymphocytapheresis, and staph protein-nA immunoadsorption as an immunomodulatory therapy in pateints with AIDS and AIDS-related conditions.' *Journal of Clinical Apheresis* 3:133–139.

Kleine, M.W., Stauder, G.M., Beese, E.W. (1995) 'The intestinal absorption of orally administered hydrolytic enzymes and their effects in the treatment of acute herpes zoster as compared with those of oral acyclovir therapy.' *Phytomedicine* 2:7–15.

Patney, N.L., and Pachori, S. (1986) 'A study of serum glycolytic enzymes and serum B hepatitis in relation to LIV.52 therapy.' *The Medicine and Surgery* 4:9.

Singh, V. K. *et al* (1997) "Circulating autoantibodies to neuronal and glial filaments in autism." *Pediatric Neurology* 17:88-90.

Uhlmann, V., and O'Leary, J.J. (2000) 'Measles virus (MV) in reactive lympho-nodular hyperplasia and ilio-colitis of children.' Paper Presented at 180th Meeting of the Pathological Society of Great Britain and Ireland. Westminster, London January 18-21.

The immune system

The National Institute of Allergy and Infectious Diseases (NIAID) supports research studies on the function of the immune system in various diseases; ways to prevent and treat autoimmune disease; studying disease pathogenesis and investigating new ways to modify the immune system. Numerous publications on many autoimmune diseases available at: www.niaid.nih.gov/publications/ Or write: NIAID, Office of Communications; Bldg. 31/Rm. 7A50; 31 Center Drive, MSC 2520; Bethesda, MD 20892–2520. Telephone: (301) 496–5717.

Sulfur and Epsom salts

Bamforth, K.J. (1997) 'Reduced sulphate conjugation and lower plasma sulphate in autistics.' *Developmental Brain Dysfunction* 10:40–43. [The inability to effectively metabolize certain compounds particularly phenolic amines, toxic for the CNS, could exacerbate the wide spectrum of autistic behavior.]

Harris, R.M., Hawker, R.J., Langman, M.J., Singh, S., and Waring, R.H. (1998) 'Inhibition of phenolsulphotransferase by salicylic acid: A possible mechanism by which aspirin may reduce carcinogenesis.' *Gut* 42(2):272–275.

Jones, A.L., Roberts, R.C., and Coughtrie, M.W. (1993) 'Common food additives are potent inhibitors of human liver 17 alpha-ethinyloestradiol and dopamine sulphotransferases.' *Biochemical Pharmacology* 46(10):1713–1720.

Total load (Effects on neurology and behavior)

Accardo, P., Whitman, B., Caul, J., and Rolfe, U. (1988) 'Autism and plumbism. A possible association.' *Clinical Pediatrics* 27(1):41–44.

Boris, M., and Mandel, F. (1994) 'Foods and additives are common causes of the attention deficit hyperactive disorder in children.' *Annals of Allergy* 72:462–468.

Cagiano, R., *et al.* (990) 'Evidence that exposure to methyl mercury during gestation induces behaioral and neurochemical changes in offspring of rats.' *Neurotoxicology and Teratology* 12(1):23–28.

Cooke, K., and Gould, M.H. (1991) 'The health effects of aluminum – A review.' *Journal of the Royal Society of Health* 111(5):163–168.

Dumont, M.P. (1989) 'Psychotoxicology: The return of the mad hatter.' *Social Science and Medicine* 29(9):1077–1082.

Goyer, R.A. (1990) 'Lead toxity: From overt to subclinical to subtle health effects.' *Environmental Health Perspectives* 86:177–81.

Halsey, N.A. (1999) 'Limiting infant exposure to thimerosal in vaccines and other sources of mercury.' *Journal of the American Medical Association* 282(18):1763–1766.

Landrigan, P.J., Graham, D.G., and Thomas, R.D. (1993) 'Strategies for the prevention of environmental neurotoxic illness.' *Environmental Research* 61(1):157–163.

McBride, W.G., Black, B.P., and English, B.J. (1982) 'Blood lead levels and behaviour of 400 preschool children.' *Medical Journal of Australia* 10;2(1):26–29.

Ross, G.H. (1997) 'Clinical characteristics of chemical sensitivity: An illustrative case history of asthma and MCS.' *Environmental Health Perspectives* 105 Supp 2:437–41.

Rossi, A.D. Ahlbom, E., Ogren, S.O., Nicotera, P., and Ceccatelli, S. (1997) 'Prenatal exposure to methyl mercury alters locomotor activity of male but not female rats.' *Experimental Brain Research* 117(3):428–436.

Ruppert, P.H., Dean, K.F., and Reiter, L.W. (1985) 'Development of locomotor activity of rat pups exposed to heavy metals.' *Toxicology and Applied Pharmacology* 78(1):69–77.

Selvin-Testa, A. (1991) 'Chronic lead exposure induces alterations on local circuit neurons.' *Microscopia Electronica y Biologia Celular* 15(1):25–39.

Tirado, V., Garcia, M.A., Moreno, J., Galeano, L.M., Lopera, F., and Franco, A. (2000) 'Neuropsychological disorders after occupational exposure to mercury vapors in El Bagre (Antioquia, Colombia).' *Revista de Neurologia* 31(8):741–742.

Trocho, C., Pardo, R., Rafecas, I., Virgili, J., Remesar, X., Fernandez-Lopez, J.A., and Alemany, M. (1998) 'Formaldehyde derived from dietary aspartame binds to tissue components in vivo.' *Life Sciences* 63(5):337-49.

Watanabe, C. (2001) 'Selenium deficiency and brain functions: The significance for methylmercury toxicity.' *Nippon Eiseigaku Zasshi* 55(4):581–589.

Food: Effects on neurology, the brain, and behavior in children

Allen, D.H., Van Nunen, S., Loblay, R., Clarke, L., and Swain, A. (1984) 'Adverse reactions to foods.' *Medical Journal of Australia* 141(5 Suppl):S37–42.

Anderson, J.A. (1995) 'Mechanisms in adverse reactions to food. The brain' *Allergy* 50(20 Suppl):78–81.

Arai, Y., Muto, H., Sano, Y., and Ito, K. (1998) 'Food and food additives hypersensitivity in adult asthmatics. III. Adverse reaction to sulfites in adult asthmatics.' *Arerugi 47*(11):1163–1167.

Augustine, G., and H. Levitan. (1980) 'Neurotransmitter release from a vertebrate neuromuscular synapse affected by a food dye.' *Science Magazine* March 28, 207:1489–1490.

Baumgaertel, A. (1999) 'Alternative and controversial treatments for attention deficit /hyperactivity disorder.' *Pediatric Clinics of North America 46*(5):977–992.

Berdonces, J.L. (2001) 'Attention deficit and infantile hyperactivity.' *Revista de Enfermeria 24*(1):11–14.

Breakey, J. (1997) 'Review: The role of diet and behaviour in childhood.' *Journal of Paediatric Child Health 33*(3):190–194.

Carter, C.M., Urbanowicz, M., Hemsley, R., Mantilla, L., Strobel, S., Graham, P.J., and Taylor, E. (1993) 'Effects of a few foods diet in attention deficit disorder.' *Archives of Disease in Childhood 69* (5):564–568.

Chudwin, D.S., Strub, M., Golden, H.E., Frey, C., Richmond, G.W., and Luskin, A.T. (1986) 'Sensitivity to non-acetylated salicylates in a patient with asthma, nasal polyps, and rheumatoid arthritis.' *Ann Allergy 57*(2):133–134.

Conners, C.K., and Goyette, G.H. (1976) 'Food additives and hyperkinesis: A controlled double-blind experiment.' *Pediatrics* August *58*(2):154-66.

Corder, E.H., and Buckley, C.E. (1995) 'Aspirin, salicylate, sulfite, and tartrazine induced bronchoconstriction. Safe doses and case definition in epidemiological studies.' *Journal of Clinical Epidemiology 48*(10):1269–1275.

Corvaglia, L., Catamo, R., Pepe, G., Lazzari, R., and Corvaglia, E. (1999) 'Depression in adult untreated celiac subjects: Diagnosis by the pediatrician.' *American Journal of Gastroenterology 94*(3):839–843.

Dickerson, J.W., and Pepler, F. (1980) 'Diet and hyperactivity.' *Journal of Human Nutrition 34*(3):167–174.

Egger, J., Graham, P.J., Soothill, J.F., Carter, C.M., and Gumley, D. (1985) 'Controlled trial of oligoantigenic treatment in the hyperkinetic syndrome.' *The Lancet* March 9; *1*(8428):540-5

Faulkner-Hogg, K.B., Selby, W.S., and Loblay, R.H. (1999) 'Dietary analysis in symptomatic patients with coeliac disease on a gluten-free diet: The role of trace amounts of gluten and non-gluten food intolerances.' *Scandinavian Journal of Gastroenterology 34*(8):784–789.

Feingold, B.F. (1979) 'Dietary management of nystagmus.' *Journal of Neural Transmission 45*(2):107–115.

Fernstrom, J.D. (2000) 'Can nutrient supplements modify brain function?' *American Journal of Clinical Nutrition 71*(6 Suppl):1669S–1673S.

Gerrard, J.W., Richardson, J.S., and Donat, J. (1994) 'Neuropharmacological evaluation of movement disorders that are adverse reactions to specific foods.' *International Journal of Neuroscience 76*(1–2):61–69.

Gould, J.M., and Saxer, J. (1982) 'Inhibition of norepinephrine uptake into synaptic vesicles by butylated hydroxytoluene.' *Biochemical and Biophysical Research Communications 106*(4):1106–1111.

Goyette, G.H., Connors, C.K., Petti, T.A., and Curtis, L.E. (1978) 'Effects of artificial colors on hyperkinetic children: A double-blind challenge study.' *Psychopharmacology Bulletin 14*(2):39–40.

Hindle, R.C., and Priest, J. (1978) 'The management of hyperkinetic children: A trial of dietary therapy.' *New Zealand Medical Journal 88*(616):43–45.

Jimenez-Aranda, G.S., Flores-Sandoval, G., Gomez-Vera, J., and Orea-Solano, M. (1996) 'Prevalence of chronic urticaria following the ingestion of food additives in a third tier hospital.' *Revista Alergia Mexico 43*(6):152–156.

Juhlin, L. (1981) 'Recurrent urticaria: Clinical investigation of 330 patients.' *British Journal of Dermatology 104*(4):369–381.

Kahl, R., and Kappus, H. (1993) 'Toxicology of the synthetic antioxidants BHA and BHT in comparison with the natural antioxidant vitamin E.' *Zeitschrift für Lebensmittel-Untersuchung und -Forschung 196*(4):329–338.

Kleijnen, J., and Knipschild, P. (1991) 'Niacin and vitamin B6 in mental functioning: A review of controlled trials in humans.' *Biological Psychiatry 29*(9):931–941.

Kurek, M. (1996) 'Pseudoallergic reactions. Intolerance to natural and synthetic food constituents masquerading as food allergy.' *Pediatria Polska 71*(9):743–752.

Maher, T.J., and Wurtman, R.J. (1987) 'Possible neurologic effects of aspartame, a widely used food additive.' *Environmental Health Perspectives 75*:53–57.

Moreno-Fuenmayor, H., Borjas, L., Arrieta, A., Valera, V., and Socorro-Candanoza, L. (1996) 'Plasma excitatory amino acids in autism.' *Investigacion Clinica 37*(2):113–128.

Niec, A.M., Frankum, B., and Talley, N.J. (1998) 'Are adverse food reactions linked to irritable bowel syndrome?' *American Journal of Gastroenterology 93*(11):2184–2190.

Nsouli, T.M., Nsouli, S.M., Linde, R.E., O'Mara, F., Scanlon, R.T., and Bellanti, J.A. (1994) 'Role of food allergy in serous otitis media.' *Annals of Allergy 73*(3):215–219.

Novembre, E., Dini, L., Bernardini, R., Resti, M., and Vierucci, A. (1992) 'Unusual reactions to food additives.' *Pediatria Medica e Chirurgica 14*(1):39–42.

Parke, D.V., and Lewis, D.F. (1992) 'Safety aspects of food preservatives.' *Food Additives and Contamimants 9*(5):561–577.

Petitpierre, M., Gumowski, P., and Girard, J.P. (1985) 'Irritable bowel syndrome and hypersensitivity to food.' *Annals of Allergy 54*(6):538–540.

Rowe, K.S. (1988) 'Synthetic food colourings and 'hyperactivity': A double-blind crossover study.' *Australian Paediatric Journal 24* (2):143–147.

Rowe, K.S, and K.J.Rowe. (1994) 'Synthetic Food Coloring and Behavior: A Dose Response Effect in a Double-Blind, Placebo-Controlled, Repeated-Measures Study.' *Journal of Pediatrics* November *135:* 691-8.

Salamy, J., Shucard, D., Alexander, H., Peterson, D., and Braud, L. (1982) 'Physiological changes in hyperactive children following the ingestion of food additives.' *International Journal of Neuroscience* May *16*(3-4):241-246.

Salzman, L.K. (1976) 'Allergy testing, psychological assessment and dietary treatment of the hyperactive child syndrome.' *Medical Journal of Australia* August 14 *2*(7):248-51.

Schapowal, A.G., Simon, H.U., and Schmitz-Schumann, M. (1995) 'Phenomenology, pathogenesis, diagnosis and treatment of aspirin-sensitive rhinosinusitis.' *Acta Otorhinolaryngologica Belgica* 49(3):235-50.

Schmidt, M.H., Mocks, P., Lay, B., Eisert, H.G., Fojkar, R., Fritz-Sigmund, D., Marcus, A., and Musaeus, B. (1997) 'Does oligoantigenic diet influence hyperactive/conduct-disordered children — a controlled trial.' *European Child and Adolescent Psychiatry* June 6(2):88-95.

Schnyder, B., and Pichler, W.J. (1999) 'Food intolerance and food allergy,' *Schweizerische Medizinische Wochenschrift* June 19 129(24):928-33.

Sinaiko, R. (1996) 'Medical Management of Attentional and Behavioral Difficulties of Childhood: Stimulant and Non-stimulant Strategies' http://www.feingold.org/review_sinaiko.html#[8].

Sloper, K.S., Wadsworth, J., and Brostoff, J. (1991) 'Children with atopic eczema. I: Clinical response to food elimination and subsequent double-blind food challenge.' *Quarterly Journal of Medicine* August 80(292):677-93.

Stokes, J.D., and Scudder, C.L. (1974) 'The effect of butylated hydroxyanisole and butylated hydroxytoluene on behavioral development of mice.' *Developmental Psychobiology* July 7(4):343-50.

Stolze, K., and Nohl, H. (1999) 'Free radical formation and erythrocyte membrane alterations during MetHb formation induced by the BHA metabolite, tert-butylhydroquinone.' *Free Radical Research* April 30(4):295-303.

Swanson, J., and Kinsbourne, M. (1980) 'Food Dyes Impair Performance of Hyperactive Children on a Laboratory Learning Test.' *Science Magazine* March 28 207:1485-7.

Uhlig, T., Merkenschlager, A., Brandmaier, R., and Egger, J. (1997) 'Topographic mapping of brain electrical activity in children with food-induced attention deficit hyperkinetic disorder.' *European Journal of Pediatrics* July 156(7):557-61.

Van Bever, H.P., Docx, M., and Stevens, W.J. (1989) 'Food and food additives in severe atopic dermatitis.' *Allergy* November 44(8):588-94.

Ward, N.I. (1997) 'Assessment of chemical factors in relation to child hyperactivity.' *Journal of Nutritional & Environmental Medicine (Abingdon)* 7(4):333-342.

Ward, N.I., Soulsbury, K.A., Zettel, V.H., Colquhoun, I.D., Bunday, S., and Barnes, B. (1990) 'The influence of the chemical additive tartrazine on the zinc status of hyperactive children: A double-blind placebo-controlled study.' *Journal of Nutritional Medicine* 1(1):51-58.

Yoshida, S., Mikami, H., Nakagawa, H., Hasegawa, H., Onuma, K., Ishizaki, Y., Shoji, T., and Amayasu, H. (1999) 'Amalgam allergy associated with exacerbation of aspirin-intolerant asthma.' *Clinical and Experimental Allergy* October 29(10):1412-4.

Zeisel, S.H. (1986) 'Dietary influences on neurotransmission.' *Advances in Pediatrics* 33:23-47.

Zoccarato, F., Pandolfo, M., Deana, R., and Alexandre, A. (1987) 'Inhibition by some phenolic antioxidants of Ca2+ uptake and neurotransmitter release from brain synaptosomes.' *Biochemical and Biophysical Research Communications* July 31 146(2):603-10.

Index

Disclaimer

The information contained herein is not intended to diagnose, treat, prevent, or cure any disease, or to provide specific medical advice. Its intention is solely to inform and to educate. If you have any questions about the relationship between nutrition, supplements, and your health, seek the advice of a qualified health practitioner. The reader and associated health professionals are responsible for evaluating the risks of any therapy reviewed in this book. Those responsible made every effort possible to thoroughly research the accuracy of the information and assume no responsibility for errors, inaccuracies, or omissions. Digestive enzymes are an unregulated dietary supplement classified as a safe food by the Food and Drug Association in the United States. Opinions and experiences contained herein do not reflect the position of any enzyme manufacturer, formulator, or distributor. The author has no financial interests with any enzyme or supplement manufacturer or distributor. The organizations and companies mentioned herein do not specifically endorse any particular therapy or product.

Cindy, Dana, Amber, Patti, Lynn, and I have contributed directly from our experiences, with names of other parents and their messages altered to protect the privacy of the families involved.

All product and company names are the properties of the respective organizations. All information related or pertaining to Houston Nutraceuticals Incorporated (HNI) is used with permission. www.houstonni.com or (866) 757-8627.

* If you sprinkle on food always wash
down w/ drink - can irritate mouth + skin
+ wipe mouth after taking enzymes

* Look up Thropps - strong protease on mild
in beginning?

Look for sulfite free papain

Get Capsules

To buy - Plant enzymes when possible

invertase (+ table sugar)
Drew -> Sucrase (complex sugars)
-> Protease (walnut allergy)

Me -> lactose (milk sugars)
Lipase - (fats in dairy products)
amylase (for wheat + broccoli)
protease (cheese)

Jeff -

Sara - > lactase - milk

* Give strong protease product gradually
increasing due to elimination of
toxins + waste + reaction
to this. Introduce a broad spectrum
product w/ lower protease for 2 weeks then
introduce the stronger product.